Democracy as the Political Empowerment of the People

Democracy as the Political Empowerment of the People

The Betrayal of an Ideal

Majid Behrouzi

LEXINGTON BOOKS

A Division of
ROWMAN & LITTLEFIELD PUBLISHERS, INC.
Lanham • Boulder • New York • Toronto • Oxford

LEXINGTON BOOKS

A division of Rowman & Littlefield Publishers, Inc.
A wholly owned subsidiary of The Rowman & Littlefield Publishing Group, Inc.
4501 Forbes Boulevard, Suite 200
Lanham, MD 20706

PO Box 317
Oxford
OX2 9RU, UK

British Library Cataloguing in Publication Information Available

Library of Congress Cataloging-in-Publication Data

Behrouzi, Majid, 1956–
    Democracy as the political empowerment of the people : The betrayal of an ideal /
Majid Behrouzi.
        p. cm.
    Includes bibliographical references and index.
    ISBN 0-7391-1025-X (alk. paper)
    1. Democracy. 2. Democracy—History. 3. Social choice. I. Title.

JC221.F69P52 2005
320'.01—dc22   2005001175

Printed in the United States of America

⊗™ The paper used in this publication meets the minimum requirements of American
National Standard for Information Sciences—Permanence of Paper for Printed Library
Materials, ANSI/NISO Z39.48–1992.

To the memory of my parents

We must recognize . . . that representative democracy has failed, both politically and juridically as well as socially . . . As a consequence, we must return to the fundamental meaning of "democracy," the power of the demos to govern itself. Just as the dictatorship of the proletariat rapidly became the dictatorship over the proletariat, so modern democracy quickly became a power exercised over the demos.

[I]n reality the people have no power. They neither make the laws nor govern.

—Jacques Ellul (1992)

[T]he re-creation of meaning [of democracy] depends on a political community's ability to recall its past, how it came to be confused about democracy, and its own standing *vis-à-vis* democracy. Only then will our political community remember what it means to conduct itself democratically.

—Russell L. Hanson (1989)

Contents

Part I

The Idea of Democracy in
Pre-Liberal and Non-Liberal Societies

Part II

The Case of the Liberal State and Liberal Democracy

Part III

The Case of the Late Liberal Democracy

Acknowledgements

I would like to express my gratitude to many individuals who helped me with the completion of this work. First and foremost, I am indebted to Professors Lesley Jacobs, Esteve Morera, and David McNally of York University (Canada), who directed the earlier stages of the development of this project and its companion volume. I am also indebted to Professor Richard Wellen of York University and Professor Richard Vernon of Western Ontario University for reading the manuscript and offering valuable criticisms and suggestions. I also would like to express my gratitude to Professor Claudio Duran of York University for interesting me in reading C. B. Macpherson. Moreover, I would like to thank Professor Charles Hersch of Cleveland State University for offering me ideas and suggesting sources for some parts of the book. Special thanks go to Professor Joseph DeMarco of Cleveland State University, who encouraged and guided me throughout the entire project.

In addition, I want to express my gratitude to Mr. Kevin Kay and Mrs. Barbara Hardiman for reading parts of the book and offering me helpful suggestions. Special thanks also go to Ms. Phyllis O'Linn, who read the entire manuscript several times and offered valuable suggestions and help. I am also thankful to Mr. Dick Wood for giving me permission to use his photo on the front cover of the book.

Lastly, I am greatly indebted to my wife Elizabeth, my family, and my friends, who patiently endured me through this project and offered me love, encouragement, and support.

Introduction

Democracy as the Political Empowerment of the People: The Betrayal of an Ideal is being published concurrently with its companion volume *Democracy as the Political Empowerment of the Citizen: Direct-Deliberative e-Democracy.*

The present volume argues that the conception of democracy that prevails in the general consciousness of the contemporary world is a distorted version of the "original" idea of democracy. Democracy originally meant "rule by the people." An important component of democracy in its original formulation was the ideal of the citizens' *direct* participation in the legislative and political decision-making process. The modern representative governments lay claim to being democratic, yet completely disregard this fundamental component of the idea. In the prevailing intellectual and political climate, the absence of the ideal of direct popular participation is often justified in terms of the presumed impracticality of the original idea in the complex conditions of the modern nation-state. The present volume goes against the current. To begin with, it argues that there exist ample historical evidence and compelling reasons for making the case that the absence of this ideal in the theory and practice of representative democracies results, in part, from conscious efforts that aim at discrediting the ideal; that there exist (and have existed in the past) powerful intellectual and political-economic forces which fully devote themselves to making sure that the original sense of the idea of democracy appears as impractical, even dangerous, and thus ensuring that it does not receive a fair hearing in the court of the public political opinion. To this end, the present volume offers a short conceptual history of the idea of democracy. The aim here is to provide an account of the efforts and the relevant historical and theoretical developments that have contributed to the "perversion" of the original idea of democracy.[1] This presentation is not a mere attempt at telling a sad story, but rather an endeavor to present a critical examination of the hitherto-existing theories and regimes of democracy. The ultimate aim of this examination is to retrieve the original idea, and thus help prepare the political-theoretical grounds for reviving the ideal of the citizens' direct participation in making the policies and laws that shape their lives. (The latter task takes up the space of the companion volume, which will be referred to from this point on as *Direct-Deliberative e-Democracy.*)

Toward this end, the present volume argues that the notion that ordinary citizens should have *direct* and substantive roles in making legislative and policy decisions is not only a deep-seated idea in the Western tradition of political thought, but also constitutes the main ideal (and a primary moral component of the idea) of democracy. Unfortunately, this ideal (and its moral substance) has been sabotaged repeatedly, as the idea of democracy has been subjected to perversion after perversion throughout its long history. The present volume further argues that, with its system of political representation, the "liberal-democratic conception of democracy" represents the latest version of these perversions.

The companion volume, on the other hand, begins where the present volume leaves off. It starts with arguing that the latest electronic technologies and media have provided the impetus for revisiting the original idea of democracy and retrieving the value inherent in the idea of the citizens' *direct* participation in politics. It also argues that these technologies and media make it possible to reformulate the idea of direct democracy in ways that would make it a workable option, and worthy of serious consideration as an alternative approach to the question of democracy in today's large nation-states. *Direct-Deliberative e-Democracy* revives and reclaims for the idea of democracy what the present volume retrieves. That is, it formulates a new theory of democracy (a theory of *e*-democracy indeed) that integrates the ideals of the citizens' direct, deliberative, and equal participation in politics into the political-cultural fabrics and institutional arrangements of present-day American liberal-democracy.

In preparing the stage for its companion volume, the present volume sets out to accomplish three goals. The first is to present an overview of the history of the perversion of the idea of democracy in the pre-twentieth-century era and, at the same time, to bring to light some of the factors that contributed to this perversion. The second goal of the present volume is to rescue the ideal of the citizens' direct participation present in the original idea from the perversions and deformations it has suffered at the hands of liberal democracy in the twentieth century. Finally, the third goal is to examine the reasons why some of the most recent theoretical attempts at retrieving the moral content of the original idea have failed. All in all, these three goals are geared toward accomplishing a major task in this volume, namely, to return to, and retrieve the original sense of the idea of democracy (and to recover the full scope of its ideals) as the idea of the *direct*, deliberative, and equal participation of all citizens in the political process, and thus lay the grounds for restoring to democracy the full scope of its ideals.

Two factors motivated the return to the original idea. The first is the desire to counter the ongoing fanfare of triumphalism—and a campaign of deception, one should add—that celebrates the existing American model of democracy as democracy *par excellence* and holds it up to the world, especially to the developing countries, as a model to emulate. The problem with the prevailing understanding of democracy, in particular in its American manifestation, is that it stands in a symbiotic relation to free-markets. Nowadays, the terms "democracy" and "free markets" appear so often together that the coupling instills in the popular consciousness of the world the false notion that democracy and free-markets are conceptually inseparable and internally connected, and that they "go together," and both can be reduced to the common category of freedom of choice. The United States sees its mission in the world as expanding "democracy and free-markets."[2] The fanfare of American triumphalism and the celebration of the victory of "democracy and free-markets" over totalitarianism in the post-Cold-War era is so loud and ubiquitous that one hardly gets to hear the view that *democracy is a moral idea, and that it essentially and conceptually has nothing to do with free-markets.* The recent popular literature, and the scholarship on democracy, by and large seem to have been mesmerized by this fanfare and succumbed to its deceptive message. Works on democracy that challenge the prevailing conception of democracy nowadays are hard to find. Democracy no longer appears as an "essentially contested concept" as it seemed a few decades earlier.[3] The free-market worshipers and their intellectual entourage seem to have finally won the contest. They have succeeded in gutting the moral substance of democracy and beating whatever is left of it into a set of anormative "rules" and a "free method" that guide the expansion of free-markets and cast an aura of legitimacy on the political universe of the market-driven societies. This is not a recent development, but rather a consummation of a massive political-intellectual undertaking that began over a century ago. By returning to the original idea of democracy, and retrieving its "true" meaning, this book intends to contribute to the undoing of this deception.

The second motivating factor for returning to the original idea of democracy is the conviction that, after twenty five hundred years, the political-moral ideals that served as the pillars of the original idea in ancient Athens still retain their appeal.[4] That is to say, they continue to command the respect of the moderns. And for this reason, these ideals can be used as a foundation for formulating a new conception of direct, deliberative, and "egalitarian" democracy that would be compatible with the realties of the contemporary world and, at the same time, could be buttressed by broader political-moral convictions and moral arguments that appeal to its citizens. This conviction also constitutes the point of departure of the companion volume and ties, as was noted above, with its claim that the latest *e*-technologies provide us with new tools to work toward realizing the old ideals in ways that the ancients themselves could not realize.[5]

In terms of architectonics, the present volume is divided into three parts. The first two parts are greatly indebted to the late C. B. Macpherson's *Real World of Democracy* and the *Life and Times of Liberal Democracy*, and follow in the tracks of the path laid by these works, albeit not in a straightforward way. Part I begins with a discussion concerning the original meaning of democracy as the idea of rule by the people, as it flourished in the golden age of ancient Athens, and then goes on to compare this idea with the muddled and distorted version of it that pervades the political consciousness of the contemporary world. This version reduces democracy to a mere "method"—and a "free" one at that—for selecting the political leaders. Part I then sets out to trace the history of the perversion and betrayal of the idea of democracy in the period that spanned from the Athenian world up to the arrival on the scene of the idea of representative democracy in the nineteenth century. This overview closely follows in the footsteps of C. B. Macpherson. It is argued in Part I that starting with the fall of Athenian democracy from glory, the meaning of the word democracy as the "rule by the (entire) people" or "power to the people" gradually degenerated to "rule only by the common people or the poor." On the one hand, this degeneration took place consciously under the aegis of the rich in an attempt to denigrate the idea of democracy. On the other hand, democracy came to represent the struggle and the reaction of the common people and the poor against the oppressive rule of the aristocrats and the rich. Moreover, it is argued that the idea of democracy in this long period was coupled with a thick notion of equality. This coupling of sovereignty and equality, it is argued, persisted through Rousseau and was finally undone in the liberal West in the nineteenth century, thus giving way to the notions of "representative democracy" and "liberal democracy." Furthermore, it is argued in this part that this de-coupling of sovereignty and equality went unheeded in the non-liberal East, and, on the strength of Marxist thought, the coupling received a new reformulation and was transformed into a new political ideal that found its highest expression in Bolshevism and the Russian Revolution of 1917. Finally, in the closing pages of Part I, it is argued that the entire range of ideals and views that came to be associated with the idea of democracy in the pre-liberal and non-liberal societies can be conceptualized as *the ideal of the political empowerment of the people.*

In continuing to trace the history of the degeneration of the idea of democracy, Part II presents a critical examination of the transformations the idea underwent in the liberal and liberal-democratic societies. It is argued here that, starting with the nineteenth century, the idea of democracy in the West first had to succumb to, and then share, the political center stage with a new and powerful political theory that appeared on the horizon in the seventeenth century. This new theory was none other than liberalism. Part II begins with a brief overview of how liberalism rose to political prominence and how the liberal state came into being. This overview also includes an account of how the modern idea of representative government was born and how it proved to be an integral part of the liberal state. Along with providing this account, a discussion of the very idea of representative government is also presented in Part II. This presentation conceptualizes the idea of representative government as resting upon five distinct political-philosophical presuppositions. This presentation is then followed by an overview of how the liberal state was transformed to a liberal-democratic one, and how, in the new society, the principles of representative government were combined with the principle of universal suffrage to produce a new conception of democracy. This new conception is referred to in this part and the rest of the book as the "liberal-democratic conception of democracy" and is characterized as the idea of "rule by a freely- and popularly-elected representative government." The last chapter of Part II briefly examines the main assumptions and premises that are often employed both in justificatory arguments for the idea of representative government and in defending this idea against the idea of the direct participation of ordinary people in governing.

The main aim of Part II is to portray the image of the "liberal-democratic conception of democracy" (and its purely representative form of government) as it is viewed from the standpoint of the ideals represented by the original idea of democracy, viz., the idea of the political empowerment of the people. The main force of the analyses, discussions, and historical data provided in this part is directed toward making three main arguments. The first is that, by design, the very idea of representative government in the liberal-democratic conception is an elitist and aristocratic construct; it is a conscious scheme designed to keep citizens at a "safe" distance away from the business of governing. The second is that the history of representative government in the modern world has adequately demonstrated that in its present form, the representative type of government works, as a rule, to the advantage of the wealthy classes (and other moneyed or well-organized interest groups) who manipulate or control the state for their own strategic purposes. These two problems constitute the main democratic shortcomings of the liberal-democratic form of representative government.[6] Finally, the third argument of Part II is that the "liberal-democratic conception of democracy"—(i.e., the idea of "rule by a freely and popularly elected representative government")—can be characterized as *the idea of the political disempowerment of the people*, and thus as a gross perversion of the original meaning of democracy as the idea of the political empowerment of the people.

Part III presents an overview of some of the criticisms and problems that the liberal-democratic state and the "liberal-democratic conception of democracy" had to fend off and grapple with in the course of the last four decades or so. Some of these criticisms target the representative component of the liberal-democratic state by questioning whether a *purely* representative system of government could truly serve the interests of the people as it claims, and whether the consent of the people is sufficient to meet the criterion of democratic legitimacy. Others argue that the democratic element in the liberal-democratic formula lacks real substance and that the liberal-democratic conception has sacrificed democracy to liberalism, as it has given primacy to choice and rights over equality and popular sovereignty. In addition to discussing the problems of the liberal-democratic state and the "liberal-democratic conception of democracy," Part III also presents a critical analysis of two alternative theories of democracy that have been developed as responses to the liberal-democratic theory in the last four decades or so—viz., the theories of participatory and deliberative democracy. This analysis shows that these new theories, though they lay bare the weaknesses of the liberal-democratic conception and offer valuable insights into how some of these problems could be remedied, by and large, they fail to overcome the democratic shortcomings of the "liberal-democratic conception" and thus fall short of restoring to democracy the full scope of its ideals.

The Conclusion revisits some of the themes discussed in Parts I-III and further develops some of the arguments presented in these parts. Moreover, the Conclusion provides a brief sketch of how the ideal of citizens' direct participation can be retrieved, and thus sets the stage for *Democracy as the Political Empowerment of the Citizen: Direct-Deliberative e-Democracy.*

Notes

1. The word "perversion" here is used in many of its ordinary senses. The eleventh edition of *Merriam-Webster's Collegiate Dictionary* defines perversion as "the action of perverting: the condition of being perverted." The dictionary defines the verb "pervert" as "1 a: to cause to turn aside or away from what is good or true or morally right: CORRUPT b: to cause to turn aside or away from what is generally done or accepted: MISDIRECT 2 a: to divert to a wrong end or purpose: MISUSE b: to twist the meaning or sense of: MIS-INTERPRET *syn* . . . debase." Moreover, throughout this work, "perversion" is often used interchangeably with words such as "distortion" and "degeneration," all of which are taken as tantamount to "betraying" the "original" ideals of democracy and twisting them into something other than what the "original" idea represented.

2. It is this conviction that informs America's understanding of globalism. In the lexicon of the American ideology of democracy, a country earns the honor of being called a "free nation" or a "democratic country" if it gives free-markets free reign, and allows them to overwhelm the political system and dictate its agenda.

3. The phrase the "essentially contested concept" is borrowed from an important essay by W. B. Gallie, titled "Essentially Contested Concepts" (written in 1956). In the essay, Gallie discusses religion, art, science, social justice, and democracy as examples of "essentially contested concepts." See Gallie (1968), pp.178-181 for his discussion of democracy. An example of the work that treated democracy as an essentially contested concept is C.B. Macpherson's *Real World of Democracy* (1965).

4. As Chapter 1 will argue, three ideals constitute the moral substance of the original idea of democracy: the citizens' direct participation, deliberation, and equality.

5. In present circumstances, marked by high levels of economic productivity, the objective conditions for democracy are much readier than they were in the time of ancient Athens. A main shortcoming of the ancient Athenian democracy was its exclusion from the franchise of the majority of the population (women, resident foreigners, and the slaves). As will be discussed later, this exclusion should, in part, be attributed to the city's under-productive economy. Without the free labor of slaves who worked in industry (mainly mining), agriculture, and in homes, without the contributions to its economy by resident foreigners, and without the tributes it collected from colonies, the city could not reach the level of the economic prosperity it needed to create the political stability, civility, and cultural-intellectual maturity necessary for establishing and sustaining democracy.

6. The sharp criticisms directed against the idea of representative government in Part II should be interpreted only as the rejection of the liberal-democratic form of the idea, and not as a wholesale rejection of the idea of political representation.

Part I

The Idea of Democracy in Pre-Liberal and Non-Liberal Societies

Chapter 1

The "Original" Idea of Democracy

Democracy came to be a popular and powerful political idea in the twentieth century. Its power and popularity gained increasing momentum as the 1980s came to a close, and finally exploded as the regime presiding over the former Soviet Union crumbled in 1991. Since then, the idea of democracy has captured the imagination of the world. Given this political power and the popularity of the idea, most regimes throughout the twentieth century, regardless of their diverse political structures and ideological orientations, declared themselves as democratic in order to claim legitimacy for their rules. "Democracy," as David Held put it in 1987, "seems to bestow an 'aura of legitimacy' on modern political life."[1] This rapid rise of democracy to the state of prominence is an amazing development if one considers the fact that throughout the history of Western civilization, until the end of the nineteenth century, democracy was generally regarded as a "bad thing."[2] This sudden reversal of attitude in Western political thought toward democracy in the twentieth century became possible in part by contorting the *original* and *literal* meaning of the idea, and in part by betraying what democracy *ideally* represents.

Literally, democracy means people power or rule by the common people. The word democracy or "*demokratia*," as Robert Dahl observes, was coined from the "Greek words *demos*, the people, and *kratos*, to rule."[3] According to Dahl, the term *demos* in ancient Athens "usually referred to the *entire* . . . people" and "sometimes . . . only [to] *the common people* or even *just the poor*."[4] Nowadays, there exists a rich literature on *who* were "the people" in ancient Athens and *how* they ruled. Briefly stated, those who ruled in Athenian democracy were free citizens of age twenty or older. The category of citizenship excluded women, immigrants, and the slaves.[5] By many indications, the people (i.e., the *free* citizens) were mainly, but not exclusively middle-class individuals—i.e., by today's definition—with varying degrees of wealth.[6] As to the question of *how* the people ruled in ancient Athens, the short answer would be that the "rule by the people" became possible via a complex political structure that centered on an "Assembly." The Assembly met the minimum of forty times a year. All citizens were eligible to attend the Assembly, *deliberate* on matters of public concern, and *directly* partake in decision makings. In the Assembly meetings, all citizens had *equal* opportunity and *equal* shares in politics, i.e., equal political power to speak, to vote on public policies, to elect, and to be elected to public service positions. The citizens were "equal before the law."[7] They directly elected officials to the key state positions in the Assembly meetings.[8] For the less important public responsibility positions, the election of officials was done by lot or rotation.[9] Aside from electing officials, the citizens in the Assembly meetings participated in deciding public policies and legislating laws. The Assembly was, in Aristotle's words, "the sovereign authority in everything, or at least [in] the most important matters, officials having no sovereign power over any, or over as few as possible."[10] The sovereignty of the demos in Athens meant that they held, as David Held puts it, "supreme authority . . . to engage in legislative and judicial functions."[11] The people exercised this authority in a *direct* manner in the Assembly.[12] They also performed executive functions in the capacity of "magistrates" when they were elected to those positions. Moreover, there were no clear lines of demarcation between the legislative and executive branches of the government in Athenian democracy.

In addition to the Assembly, there existed the Council of 500 which functioned as the executive and steering committee of the Assembly, and the Committee of 50 whose function was to "guide and make proposals to the Council."[13] The members of the Council were elected by the Assembly, and those of the Committee by the Council. The Assembly directly elected the military generals and the members of the Courts and the Councils. For the latter two bodies, the main method of election was by lot. The members of the Courts received payments for attending the sessions starting in the 460s B.C. Over the next decades, the members of the Council also received payment for their participation.[14] The quorum for Assembly was 6,000, the maximum capacity of the amphitheater. Despite the fact that the citizens—starting from 404-02 B.C.—were paid to attend the Assembly, the quorum was not achieved regularly. The actual attendance was "much less."[15] Moreover, the size of the citizen body of Athens during the time is estimated in the range of 20,000-40,000.

Furthermore, Athenian democracy reached its fully established stage around 460 B.C. with Pericles.[16] There are three important historical documents on ancient Athens' theory and practice of democracy.[17] One is the "Constitution of the Athenians" preserved in Xenophon's works. The other is Aristotle's description in his *Politics*, a part of which was quoted earlier.[18] The third is a series of oratories, the most important of all, the famous funeral speech attributed to Pericles. The speech was given at the ceremonial burial of the first Athenians who had died in the Peloponnesian war against Sparta. It is generally assumed that Pericles' speech was "'composed' by Thucydides some 30 years after its delivery."[19] In the speech, Pericles expressed the political system of his city in glowing terms as he declared:

> Our constitution is called a democracy because power is in the hands
> not of a minority but of the whole people . . . everyone is equal before
> the law: when it is a question of putting one person before another in
> positions of public responsibility, what counts is not membership of a
> particular class, but the actual ability which the man possesses. No
> one, so long as he has it in him to be of service to the state, is kept in
> political obscurity because of poverty. . . . Here each individual is in-
> terested not only in his own affairs but in the affairs of the state as
> well. . . . We do not say that a man who takes no interest in politics is
> a man who minds his own business; we say that he has no business
> here at all. We Athenians, in our own persons, take our decisions on
> policy or submit them to proper discussions . . . the worst thing is to
> rush into action before the consequences have been properly de-
> bated.[20]

This exalted description of Athenians' political philosophy, along with the brief
description of the actual workings of the city's political system provided earlier,
adequately portray what the ancient Athenians meant by democracy, and how they
practiced it with some success in the most democratic moments in their history.

Throughout most of the twentieth century, the *ideal* meaning of democracy,
as it prevailed in the popular consciousness of the world, closely followed the *literal* meaning of the word: "rule by the people." The idea that *the people* ought to
rule themselves appeared as an attractive and empowering idea to large segments
of the populations across the globe—especially to those who were living under (or
fighting against) colonial or despotic regimes. Notwithstanding this powerful presence of the ideal meaning, starting with the early to mid-twentieth century, the *real*
meaning of democracy in the West turned out to be a perverted version of the
original or literal meaning of the idea. Interpreted as the "rule by the *entire people*," the ideal of democracy in the West came to be coupled with some other political ideals that were liberal in origin, and thus had no precedent in the political
universe of ancient Athens. The most important among these were the ideals of
individual rights and liberties and the sanctity of the individual's private domains
of life. The coupling of the ideal of democracy with these liberal ideals led the
popular consciousness of the world to think of democracy as meaning "a free society." This is how the perversion of the original meaning of democracy began in
the modern West, and this is how most of the societies in the latter part of the
twentieth century came to understand the ideal represented by democracy. In the
same historical period, and parallel to this development in the West, the idea of
democracy in the East suffered a different sort of perversion. Here, being interpreted as the "rule by the *common people* or *the poor*," the idea of democracy became synonymous with "economic equality" that dominated the line of thinking
that led to two of the most important events of the first half of the twentieth century, the Russian and Chinese Revolutions.

Starting with the second half of the twentieth century, the identification of
democracy with freedom, and "democratic society" with "free society," became
widespread in the West. This perversion of the meaning of democracy in the West
was in part an ideological response to the consolidation of the Soviet regime and

the danger that the ideals of the Russian Revolution posed to the capitalist system in the West. During the Cold War, the idea of democratic society as a "free society" was promulgated as the antithesis of the totalitarian societies of the East. This propaganda campaign proved so effective that the understanding of democracy that came to dominate the general consciousness of the world could not—and still cannot—differentiate between "democracy," on the one hand, and "freedom," on the other hand.[21] Although the *ideal* of democracy is still "rule by the people," the general understanding of the idea that prevails around the globe amounts to a descriptive definition of the *practice* of democracy in Western societies. That is to say, this general understanding takes democracy as a political system in which people are "free" to "choose" their governments. Alternately stated, democracy is understood as a "free" system or "method" of electing political leaders—(which stands in direct contrast to un-free or manipulated methods of electing government officials, or having the leaders imposed upon the people). This misconception is not just a popular misunderstanding. Rather, a host of powerful intellectual, economic, and political forces have fully devoted themselves to legitimizing this misconception as the "true" meaning of democracy. Nothing serves the interests of free-markets more than reducing the idea of democracy to a category of freedom, and contorting the meaning of the "rule by the people" to that of people "making free choices."

The reasons why and how the idea of democracy was contorted into meaning a free method of electing government officials will be discussed in detail in Part II. For the time being, it suffices to state that one of the main arguments in legitimizing this perverse understanding of democracy has been the contention that the contemporary nation-state is too large and too complex to be ruled directly by the people. Since it is impractical or impossible to assemble the entire people at one place for decision-making purposes, and also given the complex nature of the issues to be decided, it is not practical, nor is it prudent, to institute a system of direct rule by the people—which was the case in the ancient city-state of Athens. Therefore, as the argument goes, the ideals of the "rule by the people" or "government *by* the people" must be abandoned and thus, must necessarily be replaced with those of the "government *of* the people" or the "government *for* the people."[22] As will be seen later, this contention is one of the main arguments that has been offered repeatedly in the course of the last one hundred years or so in support of what has been known as "representative democracy."

What is truly lost in this muddling of the meaning of democracy is the original idea of the "rule by the people." Democracy originally meant that the people were the sovereign power in the state and their sovereignty implied that they had major shares in political decision-making. That is to say, either they themselves were the *actual* political decision-makers or that they were in the position of ratifying the decisions made on their behalf by their appointees. Both the theory and practice of the representative democracy in the twentieth century went counter to this original meaning of the idea of the "rule by the people." By all indications, the opening years of the new century seem to be following in the tracks of the path laid by the previous one.

In order to find a way out of the prevailing confusion over the meaning of democracy, and as a way of attempting to retrieve the true sense of the concept—and the ideals it represents— this volume will set out to present a critical overview of the history of the idea of democracy. Starting with the following chapter, this volume will proceed to lay out and examine some of the numerous meanings and value judgments that have come to be associated with the idea of democracy in the course of its long history. As the examination will reveal, democracy has meant different things, and has represented different sets of values, for different theorists and regimes.

In what follows, this examination will use an "*idealized* conception" of the theory and practice of ancient Athenian democracy as its evaluative criteria. The conception in question will consist of two parts. The first is a normative definition of democracy that takes the "original" understanding of the concept as its "true" meaning. That is to say, it regards democracy as meaning "rule by the people," or as citizens having the *actual* and *direct* sovereign authority in the state. The second is a normative understanding of democracy in terms of a set of political-moral value-judgements that were associated with the original idea and inspired its theory and practice. These judgements can be expressed in terms of the *ideals* that can be said to have guided the theory and practice of ancient Athenians. There are three such ideals: (1) the ideal of the *direct participation* of citizens in legislative and policy decision-making, (2) the ideal of *substantive equality* (equality before the law, and equal opportunities and equal shares in contributing to decision-making), and (3) the ideal of *public deliberation* prior to decision-making.[23] In what will follow, these ideals will be taken as representing the political-moral content of the original idea of democracy. Moreover, they will be regarded as constituting an adequate set of political-moral ideals that could support a conception of direct, deliberative, and "egalitarian" democracy that would both be compatible with the realties of the contemporary world and could be justified in terms of broader moral categories and arguments that appeal to its citizens.[24] With these in mind, the ideals in question will be used in this volume as an evaluative framework in examining the various theories, and regimes, that have laid, or continue to lay, claim to standing for the ideals of democracy.

Moreover, throughout the examination, this volume will privilege the ideal of the citizens' direct participation as the "true" and "main" ideal of democracy—and its primary moral component—over the ideals of substantive equality and deliberation for two reasons. First, this ideal needs to be the focus of the retrieval project at hand, for it has suffered the most distortion in the hands of history, and yet, has received the least attention in the recent literature devoted to the aim of retrieving the moral content and the original meaning of democracy.[25] Second, there exist strong inter-links between the ideal of the citizen's direct participation in politics, on the one hand, and the ideal of the full and positive development of the human individual, on the other hand. In their inner cores, these ideals share a similar set of value-judgements—which the companion to this volume will take as its point of departure.[26] Briefly stated, these value-judgments are linked to the individual citizen's personal development and the educational utility inherent in the idea of direct participation, on the one hand, and to the potential of the idea to generate and nurture civic skills and civic virtues, and thus strengthen communal bonds, on the other hand.

It is worth adding that the "idealized conception" presented here is merely intended to draw out the *moral-ideal core* of the theory and practice of Athenian democracy, and not to romanticize or uncritically embrace this particular historical experience. One should be reminded that democracy in its birthplace was not a noble cause or a moral idea. Nor was it the genuine object or goal of Athenian politics that Pericles and others proclaimed it to be.[27] Athenian democracy, as will become evident in the rest of Part I, suffered from its own share of "democratic" shortcomings and thus failed to live up to the ideals being erected here after its experience. The legacy of Athenian democracy is a widely debated question, as scholars take stock of its triumphs and failures, examine (and re-examine) its intellectual history, and discuss how this experience could contribute to enriching our understanding of the prevailing forms of democracy.[28] It is also worth stating at this point that privileging the ideal of direct participation over those of substantive equality and public deliberation should not be interpreted as undervaluing the latter ideals. This orientation of the book follows primarily from the fact that while the ideal of direct participation has completely been ignored for the longest, the other two have either received considerable attention or won recognition, albeit in limited or distorted ways in some cases. To begin with, the ideal of substantive equality is gaining wider appeal, and in its limited sense of *formal* equality (i.e., equality before the law and the principle of "one person, one vote") has already acquired universal recognition and has become part and parcel of the contemporary world's political-moral consciousness.[29] Virtually all regimes or social-political systems, including the least democratic and most oppressive regimes, adhere to some interpretations of the principle of formal equality, albeit only nominally in some cases.[30] As to the ideal of public deliberation, there are numerous indications that the ideal is gaining ground in many social and political circles. In fact, there exists a thriving political-theoretical movement in academia in the United States that has been putting the case for deliberation eloquently for almost two decades.[31]

As the first step in starting this critical examination, beginning with Chapter 2, the rest of Part I will be devoted to presenting an overview of the meanings and value-judgements that became associated with the idea of democracy in the period that spanned the era of ancient Athens to the nineteenth century, when the regimes of representative democracy appeared on the Western political horizon. This overview will closely follow in the footsteps of Macpherson's *Real World of Democracy* and *The Life and Times of Liberal Democracy*. It will be argued that starting with the fall of Athenian democracy, the meaning of the word democracy went through a degenerative phase, in which process the meaning of the word was gradually contorted from the "rule by the *entire* people" to the "rule *only* by the *common* people or *the poor*," hence an epithet. This phase lasted until the late nineteenth century. Moreover, it will be argued that the idea of democracy in this long period was strongly linked with a substantive notion of equality. This coupling of democracy with equality was undone in the West in the late nineteenth century. The de-coupling served as the bedrock for the rehabilitation of the idea of democracy and its eventual admittance into the realm of the respectable in the twentieth century. Chapters 4 and 5 will examine the Marxian and Bolshevik approaches to the question which refused to let go of substantive equality as a main component of democracy.

Notes

1. Held (1987), p.1.
2. The phrase inside the quotation marks is borrowed from Macpherson (1965), p.1.
3. Dahl (1998), p.11, original italics.
4. Ibid., p.12, italics added. Many authors, including Carole Pateman and C. B. Macpherson, emphasize that the phrase "common people" originally meant the poor people and the word "democracy" denoted the idea of the rule of the common people over the rich elite. The following statement by Stanley I. Benn in *The Encyclopedia of Philosophy* also makes this point emphatically: "In ancient Greece, the demos was the poorer people; democracy meant rule of the poor over the rich" (Edwards 1967, Vol.2, p.338).
5. The meaning of citizenship in ancient Greece was bound with performing public responsibilities. David Held quotes Aristotle to the effect that "a citizen was someone who participated in 'giving judgment and holding office'" (Held 1987, p.36).
6. Some argue that the majority of the free citizens in ancient Athens were "workers who earned their own livings" (e.g., Jones 1964, p.18). However, despite the fact that there was a large group of poor workers ("casual labourers at the bottom"), the majority of workers can be said to have the middle-class status. With the exception of "a small group of relatively very rich men at the top" and the poor workers at the bottom, "wealth was evenly distributed, and the graduation from the affluent to the needy [was] gentle" (ibid., p.90). Starting from 404-02 B.C., the citizens were paid for attending the Assembly. The payment was meager and it is assumed that it was intended to compensate the poor workers for the time they lost from work in order to perform their citizenship responsibilities. According to A. H. M. Jones, it seems that most of the participants in the city's politics, i.e., those who attended the Assembly and juries (and ended up in the Council of 500), were of middle-class origins (and the wealthy). He deduces this "from the tone in which the orators address[ed] them" (ibid., p.50). See also the Concluding Remarks to Part I for a short discussion of the economic factors that made Athenian democracy possible.
7. Quoted from a famous oratory attributed to Pericles. (Thucydides 1954, p.117).
8. One should note that Athenians considered voting in elections as an "aristocratic rather than democratic" undertaking in that ordinary voters tended to vote for a known person rather than for an unknown figure. And this is "in point of fact that generals usually tended to be men of wealth and family" (Jones 1964, p.3, see also p.55).
9. Election by lot, unlike direct election of officials to key positions which was aristocratic in essence, was truly democratic, as it was intended to secure the principle of equality (Jones 1964, p.47).
10. Quoted from Aristotle's *Politics* by Held (1987), p.19. As Held mentions, Aristotle's account of the Athenian democracy should not be interpreted as his endorsement

of this institution. As will be seen later, Aristotle was a strong opponent of the idea of democracy.

11. Held (1987), p.17.

12. "Athenians," according to A. H. M. Jones, "attached great importance to the equality of all citizens in formulating and deciding public policy. This was secured by *the right of every citizen to speak and vote in the assembly*" (Jones 1964, p.46, emphasis added).

13. Held (1987), p.22.

14. Boardman, *et al.* (1986), p.136.

15. Budge (1996), p.25.

16. Some authors go as far back as 508-507 B.C. to mark the beginning of Athenian Democracy with Cleisthenes' reforms, e.g., Thorley (1996).

17. See Thorley (1996), pp.89-92 for a list of more ancient sources on Athenian democracy. The list also includes some modern sources.

18. Aristotle's description of Athenian democracy will be considered in further detail in Chapter 2.

19. Held (1987), p.16.

20. Thucydides (1954), pp.117-19.

21. Here the allusion is to the distinction drawn by Isaiah Berlin: "there is no necessary connexion between individual liberty and democratic rule," as he has argued in "Two Concepts of Freedom" (Berlin 1969, p.130). "The answer to the question 'Who governs me?' is logically distinct from the question 'How far does government interfere with me?'" (ibid.).

22. Here the phrases "government *of* the people," "government *for* the people," and "government *by* the people" have been taken from Abraham Lincoln's famous "Gettysburg Address." It is important to stress that Lincoln himself did *not* differentiate among these three notions. *Throughout this book, the phrase "the government by the people" will be taken to mean the direct participation of the people in decision making, whereas the phrases "government of the people" and "government for the people" will be taken to describe the circumstances wherein a select group make the decisions for the people, or in their name.* The actual quotation is "government of the people, by the people, for the people, shall not perish from the earth" (Ravitch and Thernstrom 1992, p.167). As Lummis has pointed out, this "formula . . . is taken by most people as his (for many, *the*) definition of democracy, despite the fact that he did not say it was" (Lummis 1996, p.23, original italics). Lincoln only used the "formula" to refer to American government and to "exhort . . . his audience to strive on so that the people can continue to be governed," as Lummis has argued (ibid.).

23. One should be reminded that, as A. H. M. Jones notes, "there survives no statement of democratic political theory" from ancient Greece, and all of the statements on its theory and practice, with the exception of Pericles' oratory and a few minor other sources, are "oligarchic in sympathy" (Jones 1964, pp.41-43). The two descriptions of Athenian democracy, that are generally considered as reliable, are Pericles' and Aristotle's. These two documents adequately lend support to the intuition that the three ideals listed above can be considered as guiding the theory and practice of the Athenian democracy—at least in its most democratic moments—hence the "idealized conception" proposed above.

24. This claim, and the arguments supporting it, constitute the point of departure of the companion volume. *Direct-Deliberative e-Democracy* proposes two new ideal principles for democracy and attempts to justify them by appealing to broader moral categories.

25. A critical examination of the recent literature is provided in Part III.

26. See Chapter 1 of *Direct-Deliberative e-Democracy* for a complete discussion of these value-judgements and how they relate to the ideal of the individual citizen's direct participation in politics.

27. Those who spoke of Athenian democracy in glowing terms were making virtue out of necessity. As will be seen in Chapter 2, democracy in Athens was not the "object" of Athenian politics, but rather the outcome of demos' struggle for more equality. It was a political concession offered to the demos who in seeking greater equalities "had forced their way into city politics" (Hanson 1989, p.71).

28. The reader should consult Tejera (1998), Saxonhouse (1996), Raaflaub (1983), and Claster (1967) for an overview of some of the issues involved in debates.

29. Here *formal* equality is distinguished from *substantive* equality. A discussion of the distinction between the two is offered in Part III.

30. Nevertheless, one should admit that some versions of these principles, as interpreted by some social-political systems, discriminate against oppressed minorities. Moreover, some regimes in smaller developing countries pay only lip service to these principles.

31. Chapter 13 will provide a detailed discussion of the theories of deliberative democracy which have been calling attention to the importance of deliberation to democracy since the 1980s.

Chapter 2

Macpherson and the Idea of Democracy as a "Class Affair"

By all indications, the ancient Athenian democracy in its most glorious moments was the state of rule by the *entire* people, that is to say, by all *free* citizens, regardless of their social or material status.[1] Having stated this, however, one should be quick to add that "those who dominated" this democracy, as David Held observes, "tended to be of 'high' birth or rank, and elite from wealthy and well-established families."[2] A. H. M. Jones notes that "[i]t was comparatively rare that a self-made man . . . achieved political influence," and those who did (especially the ones coming from poor or humble backgrounds) "had to face a good deal of abuse."[3] Notwithstanding their domination of the political system, the elite and the wealthy resented the sharing of power with the common people and the poor.[4] In seeking greater equalities, the demos of Athens had succeeded in forcing their way into the city's politics and the aristocrats had no choice but to accept this as a fact of political life.[5] It was precisely for this reason that democracy, in the sense of the "rule by the common people or the poor," at times was used, as Dahl puts it, "as a kind of epithet" by the city's aristocrats and their allies.[6] This usage was intended to show the aristocrats' "disdain" for the common people who had managed to break the aristocrats' previous absolute hold on the city's government.[7] This is how democracy, beginning at its very moment of birth, earned a negative connotation and became, in the words of Macpherson, a "bad word . . . a bad thing."[8] Democracy got stuck with this sort of characterization for almost another twenty-three hundred years.[9] In Macpherson's words,

> [d]emocracy originally meant rule by the common people, the plebeians. It was very much a *class affair*: It meant the sway of the lowest and largest class. That is why it was feared and rejected by men of learning, men of substance, men who valued civilized way of life.[10]

Or,

in the main Western tradition of political thought, from Plato and Aristotle down to the eighteenth and nineteenth centuries, democracy, when it was thought of at all, was defined as rule by the poor, the ignorant, and incompetent, at the expense of the leisured, civilized, propertied classes. Democracy, as seen from the upper layers of class-divided societies, meant class rule, rule by the wrong class.[11]

As one can speculate, the branding of democracy as a "bad thing" was due mainly to the fact that, until the late nineteenth century, throughout the history of Western civilization, the elite and the wealthy were in control of the cultural and intellectual productions. The phrases "the poor" and "the common man" were identical for all practical purposes. The poor could not afford culture and that is why they were the common people. The rich, on the other hand, not only had access to culture, but also were in the position of affecting all aspects of its production and development. Plato was among the first major intellectual elites who set the negative tone for democracy in the ancient world. Aristotle and others soon followed suit. Plato wrote in the period just following the fall of Athens' democracy from glory in the fourth-century B.C. As a youth, he witnessed first-hand some of the most horrific deeds of the Athenian democracy, including the call for the Peloponnesian war and the condemning of Socrates to death.

Plato was of the view that the common people lacked moral fortitude and competence, as well as the necessary technical expertise, to be in the ruling position. He characterized the "democratic type of man" as the type who gives in to his "unnecessary appetites," and thus a self-seeking and unscrupulous type. Plato attributed this in part to the excess of "liberty," and in part to the lack of education and proper upbringing and the culturalization (i.e., an aristocratic one) of those who would favor democracy. Plato's characterization of the democratic type of the state is equally bleak, and for all practical purposes can be equated with the state of lawlessness.[12] As to the need for political knowledge, wisdom, and technical expertise in the business of ruling, Plato explained this need by using the analogies of the shipmaster and the "keeper of some huge and powerful creature."[13] Plato's solution to the problem of ruling was his model of the *de jure* state, Callipolis. Callipolis was a caste system based on a strict regime of division of labor, which was designed to be ruled by an aristocracy of wise, knowledgeable, and expert elites spearheaded by a philosopher king.[14] Being the otherworldly philosopher he was, Plato's critique of democracy was not at all based on property-class relations. In fact, he opposed oligarchy more vehemently than he opposed democracy.[15] For Plato, when it came to the business of ruling, not all individuals were equally capable. Moreover, the inequality of qualifications among the most capable ones also meant that not every politically capable person could hold any office. There is a hierarchy of proficiencies and talents in Plato's theory of ruling that is part and parcel of his theory of justice in the state. In books XXXI-XXXII of *The Republic,* Plato argued that democracy is vulnerable to being hijacked by demagogues, and thus it would eventually degenerate into tyranny. Aside from his idealistic view that political decision-making must be embedded in knowledge, morality, and virtue, at the root of Plato's problem with democracy lies his deep-seated belief in what he regarded as the natural inequalities of talents among individuals and the belief that these inequalities should be mapped onto a system of political inequalities. Plato's other problem with democracy was that he held a low opinion of the "common men" and the ordinary lives they led.

The denigration of the idea of democracy continued with Aristotle. Despite offering various definitions of democracy, Aristotle took the idea to mainly represent a deviant or a "perverted" form of government where a large group (the poor) rule in accordance with their self-interests.[16] As with Plato, Aristotle too was opposed to the participation of *all* in decision-making and office-holding. However, unlike Plato who saw wisdom, technical expertise, and moral virtue as the prerequisites for participation in politics, Aristotle proposed a property-class basis for it. The common people, i.e., those who had less than a moderate amount of property, in Aristotle's view, ought not be qualified to participate. Moreover, he criticized the practice of paying citizens for attending the Assembly, "precisely because it fulfilled its purpose of enabling the poor to exercise their political rights."[17] In Macpherson's interpretation, this meant that Aristotle "was strongly opposed to full democracy: the only kind in which he found any merit was one in which 'husbandmen and those of moderate fortune' had supreme power."[18] By many indications, Aristotle's main problem with democracy sprang from his aristocratic inclinations. He believed that democracy gave too much power to the poor. The poor (the numerous) had more sovereign power than the rich (the few). Democracy was merely "the sovereignty of numbers," and not of wealth or noble-birth; nor was it the sovereignty of merit or "moral quality"—all of which Aristotle believed to originate in "culture and breeding . . . [that] . . . are more associated with the wealthier classes."[19] Aristotle's other problem with democracy stemmed from his contention that democracy allowed too much freedom to its citizens—that citizens had the freedom of "living as you like" and "doing what one likes."[20] As was the case with Plato, Aristotle equated the state of too much freedom with the state of anarchy.[21] He conceptualized the ideal state—which he called "polity"—as the one that had both oligarchic and democratic features. While the democratic features would admit large segments of the people into the franchise (mostly the middle-class farmers and the rich), the oligarchic features would make sure that not all of the people could participate in it (thus denying offices to the poor).[22]

It would be helpful to note that by the end of the fourth-century B.C., the balance of power between the rich and poor had already swung blatantly in favor of the rich, as Aristotle had preferred. Despite the fact that the Assembly continued to meet regularly, the rich increasingly exerted undue influence in the process, to the point that "[m]agistracies themselves became the preserve of the rich."[23] It is interesting to note that Aristotle, a few decades earlier, had favored this shift of power and in his *Politics* "had already offered advice to oligarchs on how to control a state."[24] The decline of Athenian democracy continued as Athens fell under the rule of Philip of Macedonia in 338 B.C., and immediately after his death, under the rule of his son Alexander the Great. With the demise of city-states under Alexander, and especially under his successor Antipater, Athenian democracy came to an end.[25]

Toward the end of the reign of Athenian democracy, a somewhat similar system of government took hold in the city of Rome. The Romans called their system *republic*.[26] The Roman Republic was first limited to the participation of the aristocrats. But, "after much struggle the common people (the *plebs*, or plebeians) also gained entry."[27] As in Athenian democracy, citizenship was restricted to indigenous adult males. The democratic participation of non-Romans in decision-making in the Roman Republic gradually expanded with its expansion of territories. Despite granting citizenship status to non-Romans, the practice of participation in politics remained confined to the city of Rome itself. This meant that those holding the honor of being the citizens of Rome had to travel to the city in order to be participants in the franchise. This inconvenience kept the overwhelming majority of Roman citizens away from participating in city assemblies.[28] With the transformation of the Roman Republic to the Roman Empire in 44 B.C., democracy perished in the territories ruled by Rome.[29] One should add in passing that the Roman Republic, as Sheldon Wolin has put it, "was not democratic in the older Greek sense, and the active conduct of affairs always remained in the hands of a relatively closed oligarchy."[30] In the words of Michael Parneti, Rome was "a republic for the few."[31] Moreover, "Roman society was sharply differentiated into several social orders [or classes]."[32] One advantage of the Roman Republic, compared with the Athenian model, was its institutional forms (elected assemblies, executive offices, courts, and the Senate) that required political activities to be conducted in accordance with pre-established institutional norms and models. This made political actions "indirect in character."[33] In the Roman Republic, there stood "between the word and the deed the distorting medium of institutions."[34] This also worked to "legitimiz[e] conflict[s]" which were unavoidable in a sharply class-divided society with conflicting interests.[35]

With the decline of the Western Roman Empire and its eventual demise in 476 A.D., and along with it, the rise of Christianity, democracy as a system of governing (or anything that remotely resembled it in the Roman Empire) became extinct. It took another six hundred years or so for the idea to begin to take hold in Europe.[36] The mediaeval world under Christianity's God-centered approach to matters of political order had no use for democracy. In his study of the state of the idea of democracy in the Middle Ages, Macpherson found no traces of either the theory or practice of democracy. What he found instead were uprisings and revolts that demanded "either a classless communistic society . . . , or a levelled society where all might have property."[37] Macpherson concluded that "[t]there is no record of any of these movements having produced any systematic theory, nor having sketched a democratic political structure."[38] It should be added that Dahl's overview of the same historical period finds some evidence to the contrary circa 1100 A.D. According to Dahl, the idea of rule by the people made a comeback in northern Italian cities such as Florence and lasted for almost two centuries. As he describes,

> participation in the governing bodies of the city-states was at first restricted to members of upper-class families: nobles, large land-owners and the like. But in time, urban residents who were lower in the socio-

economic scale began to demand the right to participate. Members of what we today call the middle classes—the newly rich, the smaller merchants and bakers, the skilled craftsmen . . . —were not only more numerous than the dominant upper classes but also capable of organizing themselves. What is more, they could threaten violent uprisings, and if need be carry them out. As a result, in many cities people like these—the *popolo*, as they were sometimes called—gained the right to participate in the government of the city.[39]

Quentin Skinner sets "the closing decades of the eleventh-century" as the beginning years of Italian city-republics and believes that "[b]y the middle of the thirteenth-century, many leading communes of Lombardy and Tuscany . . . had acquired the status of independent city-states."[40] In these republics, as Skinner states, it was understood that "all political offices should be elective" and the "right to vote was restricted to male householders, who also needed to show that they owned taxable property within city limits."[41] Despite some of their shortcomings, "the city-republics . . . developed a genuine theory of popular sovereignty; they also made serious efforts in most cases to put it into operation."[42] Skinner goes on to offer a caveat to the reader that these self-governing city-states cannot be properly regarded as "democratic" states; and given the negative connotation the word democracy received in the aftermath of the translation of Aristotle's *Politics* into Latin in the middle of the thirteenth century (i.e., "'transgressions' of good government"), these republics would not have cherished the term "democratic government."[43] That democracy was considered a "bad thing" (Macpherson's argument) in the thirteenth century is vividly evident in the following statement of Thomas Aquinas:

> a government is called democracy when it is iniquitous. . . . A democracy . . . is thus a form of popular power in which the common people, by sheer force of numbers, oppress the rich, with the result that the whole populace becomes a kind of tyrant.[44]

As to the question of why the above-mentioned democratic experiences in the medieval period do not appear in Macpherson's study, perhaps the answer would be that, as will be suggested later in this chapter, Macpherson applied the characterization of democracy as a "class affair" too strictly. Dahl's and Skinner's descriptions of these experiences suggest cases of class-compromise (between upper and middle propertied classes), whereas Macpherson's notion implies a class-warfare.

In the sixteenth and seventeenth centuries, however, Macpherson finds more explicit expressions of the idea of democracy. He mentions the democratic utopias put forth in More's *Utopia* (1516) and Winstanley's *The Law of Freedom* (1652). He also mentions English Puritanism, Presbyterians, Independents, and the Levellers.[45] Macpherson makes it clear that these movements did not put forth the ideas of "full sovereignty" or a "fully democratic franchise."[46] But rather, broadly speaking, they were more concerned with equality in property holding and the idea of "one-class society" than with the idea of rule by the people.[47]

As one moves into the eighteenth century, Macpherson's overview of the history of democracy finds more evidence of the idea. The two great events of this century, the French and American Revolutions, were influenced by a rising tide of the idea of democracy.[48] The main democratic figures in this period for Macpherson are Jean-Jacques Rousseau and Thomas Jefferson.[49] For Macpherson, Rousseau is the "classic formulator" of the "classic, pre-liberal, notion of democracy."[50] The importance of Rousseau's notion of democracy for Macpherson lies in that Rousseau relates the idea of democracy directly to the question of equality in general, and relative equality in the realm of material holdings, in particular. As Macpherson reads it, for Rousseau, democracy, i.e., the sovereignty of the "general will," is inconsistent with unlimited rights to private property.[51] For Rousseau, the sway of the general will that represents the common good would require that "no citizen shall ever be wealthy enough to buy another, and none poor enough to be forced to sell himself."[52] As Macpherson sees it, this vision of property holding in Rousseau's conception entails that the democratic system requires a "one-class society" rather than a class-divided one wherein the class-interests override the concerns with common interests.[53] According to Macpherson, the idea of a one-class society is also essential to Jefferson's notion of democracy. In Jeffersonian democracy, everyone was to be economically independent. This, as Macpherson puts it, "did not require that everyone should be a worker-proprietor, but only that everyone could be one if he wished."[54] As was the case with Rousseau's, in Jefferson's conception, democracy was a political system that admitted all adult *male* members of the society (except slaves in the case of Jefferson) and required their participation in decision-making. In the case of Jefferson, it is also worth noting here that, despite the fact that he is now widely regarded as one of the leading democrats of the American Revolution, Jefferson "did not feel comfortable with the word 'democrat'."[55] As R. Palmer has argued, the words "democrat" and "democracy" in the America of the time were regarded as "terms of abuse or reproach, smear-words, used to discredit people who would not use them of themselves."[56]

In Macpherson's overview, the pre-nineteenth century or the pre-liberal era of the conception of democracy comes to an end with Jefferson. It should be noted here that, in addition to Rousseau and Jefferson, there were other significant political thinkers and statesmen in the seventeenth and eighteenth centuries whom Macpherson considers as being "not at all democratic," for they favored class-divided conceptions of society, as well as believing in barring the non-propertied classes from participation in ruling—the most important figures falling in this category are John Locke, Edmund Burke, and James Madison.[57]

Now, as was mentioned earlier, Macpherson's overview of the history of the idea of democracy in the pre-nineteenth century era points in the direction that in this period, the idea of democracy was coupled with a strong notion of substantive equality, including equality in the realm of material holdings.[58] What really gave the idea of democracy its negative connotation in this period, in Macpherson's view, was its "leveling" and "equalizing" aspirations and implications. In Macpherson's words, "earlier models and visions of democracy were reactions against the class-divided societies of their times."[59] Moreover, according to Macpherson, the "models of democracy earlier than the nineteenth-century . . . were fitted either to a classless or to a one-class society."[60] This, in Macpherson's view, made the idea of democracy in this period essentially a "class affair."[61] Finally, it should be added that Macpherson was of the view that the thick egalitarian component of the idea of democracy, whether in the form of a one-class or classless society, was understood only as a means toward the "ultimate goal [of] . . . provid[ing] the conditions for the full and free development of the essential human capacities of all the members of the society."[62]

Having presented Macpherson's reading of the history of pre-nineteenth century democracy as a "class-affair," and acknowledging its persuasiveness, one should now point out in passing its shortcomings. First, seeing the history of democracy solely through the lens of "class affair" could lead one to overlook the democratic experiences that would not fit this model accurately. This could be one possible explanation of why Macpherson overlooked some of the democratic movements in the Middle Ages that caught the attention of Dahl and Skinner. As it appears, these movements did not involve sharp class conflicts between haves and have-nots, as was the case with the peasant revolts, and perhaps this is why Macpherson overlooked them. The second shortcoming present in Macpherson's overview lies in that his does not underscore adequately the fact, and the ideal, that the *direct participation* of (free) citizens in decision-making was a main component of the idea of democracy in this period, as was the case with ancient Athens' experience. On a related note, it is rather puzzling to learn that Macpherson believed that the idea of *democracy as the ordinary citizens' direct participation* in politics began with the New Left movement in the 1960s.[63] One wonders why Macpherson did not characterize as participatory the experience of ancient Athens or the Rousseauean and Jeffersonian models of democracy. All of these suggest that Macpherson's own conception of democracy privileged economic equality (which he regarded as the main component of substantive equality) over the value inherent in the *direct* and deliberative participation of the citizenry as the main value associated with democracy. This contention will be revisited in Part III where Macpherson's theory of participatory democracy will be examined in detail.

Returning to the eighteenth century, Chapter 3 will examine Rousseau's conception of democracy in greater detail.

Notes

1. Athenian democracy in its most glorious moments enjoyed circumstances of general economic prosperity. The importance of this favorable economic context to the success of democracy in Athens will be discussed later in the Concluding Remarks to Part I. (The reader should consult note 6 in Chapter 1 as a starter.) Moreover, one should be reminded

that males under the age of twenty, women, immigrants (foreigners), and slaves were not considered free citizens and thus were excluded from the franchise of democracy.

2. Held (1987), p.27. As A. H. M. Jones argues, there are evidence to suggest that the Council of 500 was "mainly filled by the well-to-do," and the participants in the Assembly and the Juries "consisted predominantly of middle-class citizens" (Jones 1964, p.50).

3. "The orators, who, normally holding no office, guided policy by their speeches in the assembly were also mostly well-to-do, and many of them of good family. It was comparatively rare that a self-made man . . . achieved political influence . . . poor and humbly born politicians had to face a good deal of abuse from comedians and orators" (Jones 964, p.55)

4. "There was at all times a small group of wealthy intellectuals who hated the democracy" (ibid., p.92).

5. As Hanson argues, democracy was offered as a compromise political solution by the aristocrats of the city to the demos who in seeking greater equalities "had forced their way into city politics." Democracy was "the result . . . of popular insurgency aimed at achieving greater equality" (Hanson 1989, p.71).

6. Dahl (1998), p.12.

7. Ibid.

8. Macpherson (1965), p.1.

9. The reader will encounter some examples of negative uses of the word democracy in the West later on in this chapter.

10. Macpherson (1965), p.5, italics added. The characterization of democracy by Macpherson as a "class affair" can also be found on p.12 and p.13.

11. Macpherson (1977), pp.9-10.

12. Plato's view of democracy is expressed in Chapter XXXI of *The Republic*, Plato (1945), pp.279-286. "There is," according to Plato, "so much tolerance and superiority to petty considerations; such a contempt for all those fine principles well laid down in founding our commonwealth" (ibid., p.283). Furthermore, democracy "will promote to honor anyone who merely calls himself the people's friend" (ibid.).

13. These two analogies are given in Chapter XXI of *The Republic*, Plato (1945), pp.195-96 and pp.200-01, respectively.

14. As David Held argues, in the subsequent writings such as in *The Statesman* and *The Laws*, Plato softened his critique of democracy and advocated a precursory model of "mixed state," and a "system of proportional voting which later was to find a parallel in the writings of figures like John Stuart Mill" (Held 1987, p.32). Manin expresses a similar view: "Plato's position on the subject is not reducible to the emphatic attacks expressed in the *Republic*. In the *Laws* he attempts to combine monarchy and democracy or rather, to be more precise, to find a middle way between those two forms of government" (Manin 1997, pp.35-36).

15. Plato, in *The Statesman*, according to Cornford, "regards even the more lawless type of democracy as superior to oligarchy" (Plato 1945, p.280).

16. See Aristotle's *Politics*, Book III, Chapter VII, pp.114-15. Aristotle regarded democracy as the "perversion" of the "polity" which he considered as an ideal system "when masses govern the state with a view to the common interest" (Aristotle 1958, p.114). Aristotle's various definitions of democracy are presented in Book IV, Chapter IV of the *Politics* (ibid., pp.167-69); also see Book VI, Chapters I-VII, pp.256-67, and Book III, Chapter VIII, p.116, where he defines democracy as "the constitution under which the poor, being also many in number, are in control." For a short discussion of Aristotle's

classification of the forms of government see Held (1987), pp.18-20, and Cunningham (2002), pp.6-8.

17. Jones (1964), p.49. Here, Aristotle followed in the footsteps of Plato (ibid., p.50). The practice of paying citizens for attending the Assembly was instituted in 404-02 B.C. Jones doubts that the practice was in full effect in Aristotle's time (ibid.) and notes that Aristotle was "prepared to accept political pay, provided that precautions are taken to prevent the poor outnumbering the rich" (ibid., p.145n37).

18. Macpherson (1977), p.12n1. Macpherson here bases this claim on his reading of Book IV, Chapter VI (1292b), and Book VI, Chapter IV (1318b) of an unspecified edition of *Politics*. In the latter chapter, stating his preference for an agricultural form of democracy (a form of "polity"), Aristotle proposes that "the most important offices . . . be filled by election, and confined to those who can satisfy a property qualification. The greater the importance of an office, the greater might be the property qualification required" (Aristotle 1958, p.264).

19. Ibid. The first paraphrase above is taken from Book III, Chapter VIII, p.115. See also Book VI, Chapter II, p.258 where Aristotle states: "the poor—they being in a majority, and the will of the majority being sovereign—are more sovereign than the rich." The last two paraphrases are taken from Book IV, Chapter VII, p.173, and Chapter VIII (p.175), respectively, where Aristotle considers the advantages of the aristocratic type of constitutions and the positive aspects of the aristocratic features of polity.

20. Ibid. Phrases inside the quotation marks are taken from p.258 and p.234, respectively.

21. This point is argued by Ernest Barker, the editor of *Politics*. Barker further argues that both Plato and Aristotle regarded the state of too much freedom as "the negation of the city-state." Moreover, it was their fear of anarchy—rather than their aristocratic inclinations—that made them ("Plato far more than Aristotle") dislike democracy (Aristotle 1958, pp.liii-liv). Barker goes on to argue that both Plato and Aristotle exaggerated the anarchic aspects of democracy and that "[d]emocratic government in the fourth-century did not mean anarchy" (ibid., p.liv).

22. The "polity" was an ideal system where "masses govern the state with a view to the common interest" (ibid., p.114). See also pp.263-64. For detailed descriptions of polity see Book IV, Chapters VIII-XII, pp.174-86. Moreover, as was quoted in support of Macpherson's claim above, in Book VI, Chapter IV (1318b) of *Politics*, Aristotle proposes that "the most important offices . . . be filled by election, and confined to those who can satisfy a property qualification. The greater the importance of an office, the greater might be the property qualification required" (Aristotle (1958), p.264). In this form of "polity," all free-born males (mostly farmers) are allowed to vote, hold the elected officials accountable, and form the court of law (ibid., p.263-64).

23. Boardman, *et al.* (1986), p.333.

24. Ibid. The authors quote the following statement from Aristotle's *Politics*, Book VI, Chapter VII, "Those who enter into office may also be reasonably expected to offer magnificent sacrifices and to erect some public building, so that the common people, participating in the feats and seeing their city embellished with offerings and buildings, may readily tolerate a continuation of this constitution [oligarchy]" (ibid., p.333). One should be reminded that *Politics* was written between 335 and 322 B.C., in the period after the conquest of Athens by Alexander the Great.

25. As to when the Athenian democracy came to an end, there are no definite answers or dates. While some authors declare Athenian democracy dead after Antipater revised Athens' constitution to require property qualifications for full citizenship in 322-21 B.C.

(in effect reducing the size of Assembly in half), others report that there were "democratic interludes" after 322 B.C., e.g., Sinclair (1988), p.22.

26. The phrase "republic" comes, as Dahl indicates, "from *res*, meaning thing or affair in Latin, and *Publicus*, public: loosely rendered, a republic was the thing that belonged to the people" (Dahl 1998, p.13).

27. Dahl (1998), p.13.

28. Ibid., p.14.

29. See Parenti (2003) for a description of the workings and failings of democracy in the Late Republic (133-40 B.C.), pp.49-58.

30. Wolin (1960), p.83.

31. Parenti (2003), the title of Chapter 3.

32. Wolin (1960), p.83. See Parenti (2003), pp.27-43 for a description of the class structure of Roman society and a vivid picture of the brutality and inhumanity inflicted upon the lower classes.

33. Wolin (1960), p.84.

34. Ibid.

35. Ibid.

36. Working with the fall of the Roman Republic as the beginning of the fall of democracy in the ancient world (*viz.*, Europe), Dahl states that democracy (or popular sovereignty, as he puts it), "vanished from the surface of the earth for nearly a thousand years" (Dahl 1998, p.15). It should be commented that Dahl's claim here overlooks the experiences of non-Europeans with some forms of political decision-making that shared some common features with the Western understanding of democracy. The reader should consult Sen (2003) as an introduction to the "Global Roots" of democracy.

37. Macpherson (1977), p.13. "Where feudalism prevailed, power depended on rank, whether inherited or acquired by force of arms. No popular movements, however emerged, would think that its aims could be achieved by its getting the vote. And in the nations and independent city-states of the later Middle Ages also, power was not to be sought in that way. Where voices were raised and rebellions mounted against the late medieval social order, as in the Jacquerie in Paris (1358), the uprising of the Ciompi in Florence (1378), and the Peasants' Revolt in England (1381), the demands were for the leveling of ranks, and sometimes for leveling of property, rather than for a democratic political structure. They either wanted a classless communistic society, . . . or a levelled society where all might have property" (ibid.).

38. Ibid.

39. Dahl (1998), p.15, original italics. Dahl's overview of this period also includes the political institutions of Scandinavia, e.g., those of the Vikings (ibid., pp.17-21).

40. Skinner (1992b), p.57.

41. Ibid., p.60.

42. Ibid., p.63.

43. Ibid., pp.59-60.

44 . In Aquinas' *On Princely Government*, quoted by Skinner, ibid, p.60.

45. Macpherson (1977), pp.13-15.

46. David Wootton has challenged Macpherson's claim that most Levellers were not "full democrats." Wootton has argued that the Levellers were "nearly democrats" (Wootton 1992, pp.74-75. See also Hanson (1989), pp.74-75.

47. Macpherson (1977), p.14.

48. Hanson notes that "democracy was explicitly endorsed by some of the 'Patriots' of the Dutch Revolution of 1784-7 and the Belgian Revolution of 1789-91" (Hanson 1989,

p.75). Moreover, during the politically stormy years of 1789-99, according to Palmer, "the favorable use of the word [democracy], despite its association with Robespierre, seems actually to have increased . . . , being most frequent about 1798" (Palmer 1953, p.226). "After 1799 or 1800," however, "in the Western World as a whole, the use of the word in a favorable sense undoubtedly declined" (ibid). This trend began to change with the establishment of the Democratic Party a generation later (ibid., p.226). Moreover, "the use of the word 'democracy', as a frequent symbol of political values, dates from the period of the First World War" (ibid., p.203).

49. Macpherson (1977), p.15.

50. Macpherson (1965), p.29.

51. Macpherson (1977), p.16.

52. Quoted from Bk. II, Ch. 11 of *The Social Contract* by Macpherson (1977), p.17.

53. Macpherson (1977), p.17.

54. Ibid., p.18.

55. Palmer (1953), p.207.

56. Ibid. As Palmer points out, this was despite the fact that the favorable use of the word democracy has increased during 1789-1799 in continental Europe. By many indications, this attitude toward the word democracy in America had English origins or influences. "In England and Scotland," according to Palmer, "the antidemocrats seem to have monopolized the word" (ibid, p.223). A good example here is the widely quoted utterance of Edmund Burke on democracy in *Reflections on the Revolution in France*: "A perfect democracy therefore is the most shameless thing in the world."

57. Ibid.,p.20, and p.15. The views of Edmund Burke and James Madison will be discussed in detail in Part II.

58. Here, equality in the realm of material holding must be understood in the relative sense of the absence of economic classes, or all members belonging to one-class, and not in an absolute sense of "mathematical equality of income or wealth" (Macpherson 1965, p.47).

59. Macpherson (1977), p.10. As Macpherson is quick to point out, by "class," he means class "in terms of property" (ibid., p.11).

60. Ibid., p.20, italics added.

61. E.g., Macpherson (1965), p.5, p.12, and p.13.

62. Macpherson (1965), pp.36-37. It should be added that Macpherson intends this statement to be true for "all three concepts of democracy," viz., liberal democracy, the Marxist conception, and the third-world versions which he takes to have originated in Rousseau's conception (ibid., p. 29).

63. Macpherson (1977), p.93.

Chapter 3

Rousseau and the Idea of the Sovereignty of the "General Will"

In the eyes of his supporters, Rousseau is the greatest democratic thinker of the eighteenth century. Rousseau's ideas on political participation weighed heavily in the theoretical framework of what came to be known as "participatory democracy" in the 1960s and 1970s.[1] Contrary to this democratic appraisal of him by his supporters, in the eyes of his critics, Rousseau is the forerunner of the modern totalitarianism. Still, in the eyes of some others, parts of his writings lend support for branding him an extreme and elitist individualist.[2]

The cornerstone of Rousseau's theory of government and political legitimacy is the idea of the "general will." This notion is notoriously obscure and for this reason lies at the center of the controversy over how Rousseau should be viewed. Perhaps the best way to present Rousseau's understanding of this notion is to characterize it as being, or representing, the will of the entire community regarding the community as a single entity. Understood as such, the general will should be interpreted as representing the general (or generalizable and common) interests of the *entire* population—and not only of the majority. Thus interpreted, it then follows that the general will is to be understood as a well-intentioned will which "always tends toward the public utility" and equality.[3] An extreme stretch of this interpretation leads to the contention (Rousseau's own) that "the general will is always right," albeit "the judgement that guides it is not always enlightened."[4] It is these attributes of the general will that elevate the notion to the position of sovereignty in the state.[5] According to Rousseau, "the general will . . . direct[s] the forces of the state according to the purpose for which it was instituted, which is the common good."[6] For Rousseau, according to one interpretation, the general will exists as a "matter of fact"; it is a "factual" matter and its discovery is simply a question of finding the truth and not necessarily realizing some ideals of fairness or justice.[7] Moreover, the general will is assumed to be revealed or discovered through the mechanism of majority-rule voting.[8] In Rousseau's eyes, for a will to qualify as the general will, "it need not always be unanimous."[9] Nonetheless, "it is necessary for all the votes to be counted."[10] For Rousseau, there is a significant difference between the "general will," as defined, on the one hand, and the "will of all," on the other hand. The latter will is the sum total of the private wills of the citizens. One way to relate the "general will" to the "will of all" or the "private wills" is to argue that the latter wills, taken individually, all have a common component that can be laid bare once their particular components—that "tend toward having preferences"—have been cast aside.[11] This common component of the private wills constitutes the general will.[12] Another way of relating the general will to the will of all is to say that the general will, as Rousseau sees it, is the "constant will of all the members of the state," an enduring will that represents the enduring interests of all members vis-à-vis the transient particular or private interests of the individual members aggregated in the will of all.[13]

As to the question of sovereignty, Rousseau's conception is as obscure as the general will. He regards sovereignty as being "merely the exercise of the general will."[14] As he puts it, "only the general will can direct the force of the state."[15] For Rousseau, sovereignty is both the authority to make laws *directly*, and the authority to have them executed or enforced, preferably *indirectly*, as will be discussed later. The main sense of sovereignty in Rousseau is the legislative authority. Rousseau is very clear on the question of representation in legislative affairs: "the sovereign . . . cannot be represented by anything but itself" or, "the people cannot be represented in the legislative power."[16] However, one should add that Rousseau also seems to accept the proposition that the appointees or deputies of the people could do the legislating for them, provided that the people get the opportunity to review the laws the appointees have drafted and *ratify* them "*in person.*" This claim is inferred from a famous and widely quoted passage in *The Social Contract* where Rousseau indicts the British representative system of government as a form of "slavery" and claims that "[a]ny law that the populace has not ratified in person is null: it is not a law at all."[17] Regarding its authority to execute or enforce the law, the sovereign can either execute laws directly or elect/appoint officials to execute them on its behalf—Rousseau favors the latter option.[18] Furthermore, Rousseau maintains that sovereignty "cannot be alienated."[19] He goes on to argue that sovereignty is also "indivisible for the same reason that it is inalienable."[20] It should be added that by claiming the indivisibility of sovereignty, Rousseau mainly means to rebuke those political theorists who in their advocacy of the division of the state powers into legislative and executive, or into internal affairs and external affairs, lose precision in what they advocate, and end up confusing laws— (which are supposed to be *general* in object and source)—with their particular applications, e.g., a particular policy.[21]

Having presented these brief exposés of Rousseau's understanding of the general will and sovereignty, one might ask, how do these conceptions fit within the overall system of the political thought of Rousseau? The best place to start to sketch a general picture of Rousseau's system would be his version of the social contract theory. Here, the natural goodness of Man and his freedom is a given for Rousseau. In his natural state, Man is a free (self-mastery), solitary (asocial), self-regarding (self-love), and self-sufficient being. Moreover, he is an amoral and arational being; yet he has a capacity for morality, reason, and especially for compassion. However, the circumstances of his life such as the desire for companionship and satisfying his natural needs force him to live in the company of others, hence social life or society. The price he pays for this is hefty. Society not only enslaves him—("man is born free but everywhere he is in chains")—but also corrupts his nature.[22] The interdependence society creates among its members makes them evil by compelling them to manipulate one another and seek the exploitation of their fellow citizens. Thus, in a radical break with the past, Rousseau identifies society, and not nature, nor the supernatural, as the source of evil in human life. This suggestion, necessarily, leads Rousseau to put forth an equally radical proposal for overcoming the evil in Man's social life, viz., the drafting of a new social contract that reconstructs society in ways that it would closely mirror the relationship between Man and nature in the state of nature, and with it the creation of a new form of government designed specifically for the purpose of ameliorating the conditions of Man's existence in the state of society. Such a state would advance the common interests of all and simultaneously keep in check the selfish needs and wants of individuals by establishing laws that would be "impersonal," "objective," "general in their applicability," and egalitarian at the same time.[23] What would persuade individuals to consent to the new social contract would be their acquired reason and the common components of their self-interests.

However, knowing that the new state itself could be a main source of evil in society—as is the case with the state of the existing social contract—and in order to ensure that it would function as it is intended, it is necessary that the power of the citizens in the new state be maximized. Thus the political power in the *de jure* state would be placed in the citizens' hands by making their common will, i.e., the "general will" the sovereign in the state. The general will is that which is decided by the citizen body.[24] This means that laws will be made *directly* by the citizens (or "ratified" by them "in person" if laws are drafted by their agents). Laws, thus made, would be the rule of all over each, as well as being binding on all citizens equally. The executive power, on the other hand, would be best if placed in the hands of a small elected aristocracy that would merely enforce the laws made by the sovereign, the people. The executive aristocracy cannot make laws because it cannot represent the people or their general will. It can only execute or enforce the laws made by them.[25] Moreover, the *de jure* state would also need to get involved, though indirectly, in laying down the foundations of morality, civic virtue, civic religion, and public education in order to assure a strong social bond for it to function properly. The *de jure* state would be founded, it appears, on a set of fundamental laws that would be laid down by a supreme grand lawgiver and then would be ratified by the entire population. Finally, Rousseau's *de jure* state is conceptualized as a small agrarian society wherein all members are equal before the law, in political power and in social status, as well as belonging to the same economic class. As to the question of economic equality, as was mentioned earlier, Rousseau believed that "no citizen should be so rich as to be capable of buying another citizen, and none so poor that he is forced to sell himself."[26] For Rousseau, the need for equality arises because "liberty cannot subsist without it."[27]

In light of this brief presentation of Rousseau's overall system of political thought, Rousseau's *de jure* state would appear to his sympathetic readers as a *state of the political empowerment of the people*, in that in this state, it is the general (and dominant) will of the people that would be placed in the position of sovereignty—i.e., in the position of authority to make laws and the authority to direct their execution. For Rousseau, the *people,* taken as individuals, should be regarded as *citizens* in that they participate in the sovereign authority.[28] At the same time, they should be regarded as the *subjects* of the state for they must obey the laws they make.[29] Interpreted as such, Rousseau's *de jure* state would literally be a democratic state. Having argued this point, one should be quick to state that Rousseau himself did not regard this system of governing as a democratic one. In his view, democracy was an impractical and utopian idea.[30] As he himself puts it, "there is no such thing as true democracy."[31] For Rousseau, the term democracy meant that the ordinary citizens would be placed in executive positions in addition to legislative ones. This meant placing the executive functions of the government in the hands of "the entire people or to the majority of the people."[32] This seems to have appeared to Rousseau as an odd thing to do, which meant there would be "more citizens who are magistrates than who are ordinary private citizens."[33] Rousseau found faults with Athenian democracy for not being able to draw clear lines of demarcation between the executive and legislative functions of governing. Lack of such a division between these functions, he believed, would create instability.[34]

There are a number of passages in Rousseau that appear disturbing to democratic minds that have developed their molds in liberal society. To these minds, to begin with, the power of the general will or the sovereign seems to be both overreaching and *absolute,* as was the case with Hobbes' Leviathan.[35] Moreover, passages such as "whoever refuses to obey the general will will be forced to do so by the entire body . . . he will be forced to be free," and "civil liberty . . . is limited by the general will," and that "the general will always is right" appear alarming against the backdrop of the twentieth-century experiences with fascism and Stalinism.[36] What does it mean to say that the general will is always right?[37] What are the implications of this statement on civil liberties? If I ended up being on the losing side of an issue that was decided by a vote, does it mean that my view was wrong? What if I insist that not I, but the winning side is wrong? Will I be forced to admit that I was wrong? Will my civil liberties be restricted by the general will if I refuse to acknowledge that I was wrong?

Of course, Rousseau was not a liberal democrat. Liberal democracy did not come into being until a century later; albeit liberalism as a political doctrine was in existence at the time and was on its way to taking shape as a state in England.[38] Rousseau was disillusioned with what he considered to be the corrupting and negative aspects of the Enlightenment. He was more interested in expanding social bond, social cooperation, and equality than in defending or promoting individual liberty. Having mentioned this, one should add that Rousseau did not lose sight of the importance of liberty. He claimed that liberty and equality were the two "greatest good[s] of all . . . which should be the purpose of every system of legislation."[39] But, he seemed to assign much greater importance to equality, "because liberty cannot subsist without it."[40] Equality, for Rousseau, was a fundamental principle of the *de jure* state. In contrast to the institutionalized inequalities of the *de facto* state which was at the root of its moral depravity, and unfreedom, as Rousseau believed, the *de jure* state would ameliorate the situation by instituting a substantive principle of equality.

Moreover, Rousseau's understanding of liberty, as is the case with those of other pre-liberal or non-liberal conceptions, lacks a recognition of individual rights. His conception of liberty should be understood more in the republican sense of positive liberties (that is to say, liberty to participate in decision-making in matters of the public good) and less in the liberal sense of protecting the individual's rights and civil liberties. In the transition from the state of nature to the state of society, according to Rousseau, Man "loses" his "natural liberties" and "gains" his "civil liberties."[41] The transition obligates Man to give up his instincts, "physical impulses," "appetite," and "possessions" and embrace or gain "justice," "moral quality," "duty," "rights," and "property."[42] Rousseau claims that civil liberty is "limited by the general will" and comes very close to arguing that it is also defined by this will.[43] This would mean that the entitlement to civil liberties would not be included as an inalienable, or self-evident, or natural, set of rights in the main body of the laws in the social-contract that would serve as the founding document of the *de jure* state. Rather, the scope and content of these rights would be pending the general will's definition or approval of these rights. Rousseau, in the words of David Held, "posited no limits to the reach of the decisions of a democratic majority."[44] This indeed appears to the present-day liberal-democratic mind as a disturbing and frightening proposition.

Furthermore, Rousseau anticipates Kant when he claims that "obedience to the law one has prescribed for oneself is liberty." Given that the general will is the will of the individual himself—(because he has participated in revealing or discovering it and he "wants" it, and it has his interests at heart after all (whether he is aware of it or not), he would not be free without it, and also because he is a member of the sovereign body that exercises it)—he would find his liberty in obeying it.[45] Thus, in obeying the general will, the individual obeys himself. And if he refuses to obey it, that is if he refuses to be free, he will be forced into being free. Despite his willingness to compromise the individual's civil liberties, Rousseau is staunchly opposed to sacrificing his equal claims to political participation and his equal rights to life and property to the common good of all. Rousseau would also reject the contention that the general will could prove antithetical to these goods as ludicrous, for the general will for Rousseau is the embodiment of them all. At the same time, however, Rousseau would be more than willing to sacrifice the private wills of the individual to the general will, and compromising his civil liberties, provided that their equal basis has not eroded. One can argue that the main idea behind developing a comprehensive system of public education, civic religion, and civic morality or virtue is that Rousseau wants to persuade the individual to do these acts of sacrifice himself, either by superseding his private wills voluntarily in favor of the general will, or by altering them so that they do not stand in its way.[46] Thus, in essence, Rousseau's *de jure* state is a moral state designed to transform the corrupted private individual of the *de facto* state to a moral and socially responsible public person, i.e., to a citizen. Despite the various restrictions on his civil liberties that he might suffer (and he would be willing to accept the suffering) the individual in the *de jure* state will be free; for he himself has willed these restrictions and he is part and parcel of the sovereign body that has drafted them and is in charge of enforcing them.

These transforming tasks assigned to the *de jure* state are enormously profound and overarching. Yet by and large, they are envisioned to be carried out indirectly by the leading figure of the *de jure* state. The founding father of the *de jure* state, "the legislator," is a "charismatic" leader, and an "extraordinary man," with "superior intelligence."[47] He is a god-like and mythical figure with superhuman powers.[48] Moreover, he is neither a magistrate nor a sovereign, and his office "which constitutes the republic, does not enter into its constitution."[49] He frames the law, yet does not have any legislative rights; nor does he exercise any authority over the citizens.[50] Furthermore, he is "incapable of using either force or reason," yet he can "compel without violence and persuade without convincing."[51] As Dante Germino has suggested, "the term ['the legislator'] is a misnomer."[52] This is because, "[t]he legislator . . . does not make the laws, but rather proposes the initial, fundamental laws to the people . . . , the people as a whole remain the true 'legislator'."[53] For Rousseau, "the legislator" enlightens the people and guides their judgement, thus setting in motion the process of the formation of the general will that will be revealed by a public vote. In order for the people to be enlightened, "persuaded," and "compelled" by a man who is "incapable of using either force or reason," one should naturally assume that they must believe in a system of myths or religion. From this point on in *The Social Contract*, Rousseau slides down a slippery slope of wedding politics to myths and religion.[54] Rousseau's *de jure* state, as Andrew Levine puts it, "requires legitimating myths. . . . In this respect, Rousseau stands in the venerable Platonic tradition of the 'golden lie'."[55]

Finally, it is worth reiterating here that the idea of the *citizens' direct participation in legislative functions of the state* is an essential component of Rousseau's political thought. Rousseau saw the value inherent in the idea of the citizens' direct participation in exercising the general will (i.e., in taking part in sovereignty) primarily in terms of its regulatory effects. Direct participation would keep the elected executive aristocracy in check. As will be discussed later, the idea of "participatory democracy" that thrived in the 1960s and the 1970s took a serious reading of Rousseau's works. As will be seen in Part III, Rousseau's theory constituted the main pillar of Carole Pateman's theory of participatory democracy in her *Participation and Democratic Theory*. Pateman, a major theorist of the idea of participatory democracy, emphasized that the value of participation in Rousseau lay mainly in that it gave the individual "a very real degree of *control* over the course of his life and the structure of his environment."[56] This gave the individual "actual, as well as . . . [a] . . . sense of, freedom."[57] Moreover, in Pateman's view, "a second function of participation in Rousseau's theory is that it enables collective decisions to be more easily accepted by the individual."[58] The "integrative function" of participation is a third aspect of participation in Rousseau's theory, according to Pateman: "it increases the feeling among individual citizens that they 'belong' in their community."[59] Based on this reading of Rousseau, Pateman believed that for Rousseau, "the major function of participation is an educative one."[60]

Pateman's reading of the notion of participation in legislation in Rousseau puts it in a positive light. But one should add that there also exists a negative side to this notion. That is to say, one can argue that the notion of participation in Rousseau can be characterized as a subtle case of rubber stamping of what is intimated to the public as the general will by the supreme legislator who camouflages what he intimated to people as the enlightenment of the public. In this negative reading of participation, "the legislator" turns out to be also a grand manipulator. This negative characterization is warranted in light of Rousseau's suspicion of public deliberations and the importance he assigns to the enlightening role of the legislator.

On the question of public deliberations, Rousseau saw no reason for "a sufficiently informed populace" to have "communication among themselves" when they assemble, or as he puts it, when they "deliberate."[61] It seems that for Rousseau, deliberation means nothing more than an occasion when citizens come together to acquire information (or get enlightened), or to cast their votes—or perhaps this is the only meaning of deliberation that he is willing to allow.[62] By all indications, Rousseau was concerned that the truth-finding aspects of public deliberations, and factionalism or factions (or as he puts it, "intrigues and partial associations") that would develop in the process, would get in the way of the task of revealing or discovering the general will.[63] Rousseau believed that "the populace always wants the good, but by itself it does not always see it."[64] Moreover, he was of the view that the populace is "often tricked."[65] Thus, "the judgement that guides it [the good, the general will] is not always enlightened."[66] That deliberations could result in the citizens being deceived and thus being led to follow demagogues seems to be a deep-seated fear of Rousseau. And this seems to be what Rousseau believed was wrong with Athenian democracy. According to Joseph Femia, "Rousseau had little patience with those who exalted the deliberative virtues of Athenian democracy."[67] He quotes Rousseau's view on Athens: "for Athens was in fact not a democracy, but a very tyrannical Aristocracy, governed by philosophers and orators."[68]

If deliberations (understood as forums for the exchange of ideas among the citizens) and "partial associations" (understood as opposing and competing conceptions of the good of the community) are not allowed, then how can the citizens be expected to be "sufficiently informed" or to be able to develop an understanding of what the general will on the matter put to vote would or should be? Here, "the legislator" enters the equation. One is justified in suspecting that in the absence of opposing views, the legislator would have the clout and prestige to hint at what the general will would or should be. This is what Germino seems to be suggesting in the following passage:

> [t]he general will [for Rousseau] does not automatically emerge from
> the deliberations of a people, for individuals may be seduced by the
> temptation to accord primacy to their own interests. And even if this
> temptation is overcome and the people called into assembly conscien-
> tiously seek the general will, their judgment about it may be mistaken.
> Consequently, the judgment of the people is in need of "enlighten-
> ment." From such "public enlightenment," from the "union of under-
> standing and will," come decisions in conformity with the general will.
> But such illumination of the collective judgement of a people will take
> place only under the proper guidance. This is why a legislator is neces-
> sary.[69]

Moreover, given that the citizens in the Rousseauean *de jure* state believe in
myths, one should also wonder whether they would be able to reason their way on
their own to discovering the truth, that is to say, to formulating the general will.
And this is another reason why the figure of a supreme leader ("the legislator") is
indispensable to the Rousseauean *de jure* state. Such an understanding of how the
general will is assumed to be revealed, no doubt forces one to conclude that the
notion of the participation of citizens in decision-making in Rousseau is tainted for
being overshadowed by the figure of "the legislator."

     *   *   *   *   *

In order to recover the democratic core of Rousseau's conception of sovereignty—
and to bring it to conformity with the original idea of "rule by the people"—one
needs to cleanse the conception from contortions that the powerful presence of the
legislator introduces into it. In order to do this in a theoretically coherent fashion,
one needs to return to the central idea of the "general will" and begin with abandoning Rousseau's extreme interpretation of what it represents and, at the same
time, reject his understanding of how the will is assumed to be revealed. This in
turn would require abandoning altogether the claim that the general will exists as
an objective "matter of fact" waiting to be revealed and, at the same time, it would
require qualifying the claim that "the general will is always right." Doing so would
leave one with a thinner notion of the "general will" that would treat it exclusively
as a will that stands for the common or generalizable interests of the community
and its ethical ideals. This would accord with Rousseau's main characterization of
the general will, that is, a will that "tends toward the public utility" and equality.
As construed, one would then be able to argue that the general will could lay claim
to being "always right" only in that it has the realization of the generalizable interests of the public and its ethical ideals (most important of all, equality) as its object
and focus, and such a will can never be regarded as wrong by any "virtuous" citizen or any unified citizen body (composed primarily of virtuous and civic-minded
members). It should be noted that the general will in this thin version could also be
regarded as the "popular will." What makes it popular is that it is generally willed
by individual citizens, for it aims at realizing the generalizable interests of all. Although this interpretation of the general will satisfies, in a narrow sense, Rous-

seau's claim that the general will is always right, it does not reveal anything about the content of the interests it represents, nor does it make any claims as to how these interests can be known, let alone lend credence to the extreme claim that the general will is an objectively valid matter of fact, or a "rational will."

Alternately stated, this thin notion takes the general will primarily as a "popular will" which is generally willed by a competent citizen body. Viewed as such, the "general will" draws its political legitimacy or its claim to "righteousness," not from the claim that it is objectively valid or rational (i.e., the claim that it is derived from expert knowledge and wisdom and/or that it accords with universal reason, e.g., the eternal truths such as justice, equality, and freedom), but from the empirically verifiable claim that it is generally (i.e., popularly) willed by a citizen body composed of civic-minded members. This interpretation, as will be seen in what follows, opens the door to the possibility that the general will could also lay claim to being rational, or at least, to being "reasonable."

That Rousseau's main purpose in *The Social Contract* is to reach, in the general will, a "synthesis of rational will and popular will" is a point argued repeatedly by Jason Neidleman.[70] Neidleman contends that Rousseau does not reach his goal completely, and the general will has to settle for "a state of perpetual striving . . . for a reconciliation of popular will and rational will."[71] Neidleman characterizes this striving as an "activity," which he takes as constituting the essence of what he calls "citizenship"—hence the title of his book on Rousseau: *The General Will Is Citizenship*.[72] Using Neidleman's contention, one can argue that the general will in the thin version adopted earlier could be viewed as a popular will that "strives" to be also rational, or at least, "reasonable"—and could succeed in this endeavor in ways that Rousseau's extreme version could not.[73] To make this argument, one needs to further assume that the "virtuous" citizens, by and large, are also politically knowledgeable. Moreover, one also needs to assume that they (or at least the majority of them) possess sufficient levels of cognitive competence and reasoning skills that would enable them to reflect upon the community's generalizable interests, while keeping an eye on its ethical standards. By granting these assumptions, one can achieve a synthesis of the popular will and the rational will that remained out of the reach of Rousseau's own interpretation of the general will. And through this synthesis, one can let Rousseau's conception of sovereignty reveal its democratic core. Thus, in a democratic *de jure* state, in the "activity" of "citizenship," the civic-minded, politically knowledgeable, reasonable, and cognitively competent citizens would bring their (differing) formulations of the generalizable interests and their (differing) interpretations of the ethical ideals of their community to the attention of their fellow citizens and subject them to the scrutiny of public reason and, eventually in the process, would manage to compose a will that would be ratified in a public vote and thus would be crowned as the "general will."[74]

This discussion of the thinner version of the general will and the contention that it could be both popular and reasonable at the same time leads directly to arguing that one's desire to recover the democratic core of Rousseau's conception of sovereignty would compel him to also reject Rousseau's anti-deliberative stand,

and with that, to reject the monopoly of the legislator as the sole authority on the questions of public education, political knowledge, and wisdom. The problem with Rousseau's prohibition of public deliberations is not just that it subverts the meaning of the political (as it outlaws factions and struggles among competing conceptions of the common good). The bigger problem has to do with the fact that this prohibition denies to citizens the right and access to the public-dialogical methods and means of receiving political education. It also denies to them the access to public (i.e., openly and amply provided) forms and sources of social-political knowledge. Finally, this prohibition robs citizens of the educational opportunities they need for developing their cognitive and reasoning skills and other competencies that are of great utility in formulating (or "discovering") the general will.[75]

Thus, by refusing to grant to "the legislator" the dominant role as the political educator and the sole authority on the questions of social-political knowledge and wisdom, and in the same vein, by abandoning Rousseau's ban on public deliberation, one is left with what is truly democratic in Rousseau's conceptions of sovereignty and participation.[76] That is, concisely stated, the idea that the people ought to be the authors of the laws they are expected to abide by; that sovereignty is the collective exercise of the power of decision-making by the people; that this sovereignty is directed by the force of the general will (the thin version) which is formed collectively (and deliberatively) by the people around the question of what constitutes their generalizable interests and their political moral ideals; that collective decision-making must focus on realizing these interests and ideals; and finally, the idea that sovereignty cannot be represented.

Notes

1. Some believe that Rousseau's ideas also influenced the theories of deliberative democracy—e.g., Joseph Femia speaks of Rousseau as "a messiah of deliberative democracy" (Femia 1996, p.389).

2. This individualistic side of Rousseau is present in his latest works such as the *Confessions* and the *Reveries of a Solitary Walker*. In addition to the *de jure* state that is intended to make human beings "fully united," the other possible Rousseauean solution to the problem of human existence would be to make them "wholly separated" and independent. This reading of Rousseau is suggested by Arthur M. Melzer in *The Encyclopedia of Democracy* (Lipset 1995, Vol.3, p.1089) and in Melzer (1990), e.g., pp.89-94. The elitist component of this individualism lies in that, according to Melzer, "this solution is not practicable for the vast majority but works only for the rare, gifted individual" (Lipset 1995, Vol.3, p.1089).

3. Rousseau (19871762], p.31. The full statement is as follows: "It follows from what has preceded that the general will . . . always tends toward the public utility" (ibid.). Moreover, "the general will tends toward equality" (ibid., p.30).

4. Ibid., p.38, also p.31.

5. Ibid., "sovereignty is merely the exercise of the general will" (ibid., p.29).

6. Ibid., p.29.

7. This is Andrew Levine's reading of *The Social Contract* (Levine 1976, p.56).

8. Levine expresses the view that "Rousseau is a majority-rule democrat, to be sure; but he is not dogmatically committed to simple majority-rule voting, except as a principle. He is willing to countenance and sometimes even to accede to conservative arguments for larger majorities" (Levine 1976, p.62).

9. Rousseau (1987 [1762]), p.30.

10. Ibid.

11. Ibid., pp.29-30.

12. According to Rousseau, "[t]here is often a great deal of difference between the will of all and the general will. The latter considers only the general interests, whereas the former considers private interest and is merely the sum of private wills. But remove from these same wills the pluses and minuses that cancel each other out, and what remains as the sum of the differences is the general will" (Rousseau 1987[1762], pp.31-32). Andrew Levine's interpretation of this obscure relationship between the general will and the private will in Rousseau is that "the general will is and is not a private will" (Levine 1976, p.43).

13. Rousseau (1987 [1762]), p.82.

14. Ibid., p.29.

15. Ibid.

16. Ibid., p.29 and p.75, respectively.

17. Ibid., p.74. Rousseau's exact words in indicting the English representative system are as follows. "The English people believes itself to be free. It is greatly mistaken; it is free only during the election of the members of Parliament. Once they are elected, the populace is enslaved" (ibid.). However, in an obscure work titled *Considerations on the Government of Poland* that was completed in 1772 (ten years after the publication of *The Social Contract*), but was not intended for publication, Rousseau took exception and accepted a delegatory form of government for Poland. It is worth noting that Rousseau's acceptance of this form of government was motivated by practical considerations in the specific circumstances of Poland at the time. Moreover, Rousseau required that the deputies be instructed by their constituents regularly and be bound by their instructions. In addition, he requires that the deputies be changed or reelected frequently in order to ascertain that they would not lose touch with "the will of nation" (Rousseau's *Political Writings* (1986), pp.187-205). It should be mentioned that in the same work, Rousseau expresses the same negative view about the English system as he did ten years earlier in *The Social Contract*: England "has lost her liberty" for she has neglected the principle of *frequent re-electing* of deputies (ibid.,p.188).

18. For Rousseau," the people cannot be represented in the legislative power. But it can and should be represented in the executive power, which is merely force applied to law" (Rousseau 1987 [1762], p.75).

19. Ibid., p.29. Also, "the people cannot be represented in the legislative power" (ibid., p.75). Moreover, one can allude to his famous statement where he claims that the English people are mistaken to think that they are free because of their representative system of government (ibid., p.74).

20. Ibid., p.30. Here, Rousseau rejects and affirms Hobbes at the same time. He argues for the indivisibility of the sovereign as Hobbes did (e.g., in Hobbes 1996 [1651], Chapter XIX, p.130) while completely rejecting Hobbes' notion of fully alienated sovereignty.

21. See Rousseau (1987[1762]), pp.30-32. Here Locke and Montesquieu come to mind even though Rousseau does not mention them by name.

22. Rousseau (1987 [1762]), p.17. That Man is good by nature but corrupted by society is one dominant theme of *Emile*. Rousseau's *Second Discourse* tells the story of the fall of Man.

23. This interpretation of Rousseau's understanding of Man in the state of nature and his proposed social contract is taken from Melzer (1990) and Porter (1986). According to Porter, Rousseau believed that the most important features of the relationship between Man and nature were present in natural laws and consisted of "impersonal," "objective," and "general in their applicability," all of which were "based on equality" (p.55). Rousseau believed that social laws should mirror these three features of natural laws (p.53).

24. "The sovereign, having no other force than legislative power, acts only through the laws. And since the laws are only authentic acts of the general will, the sovereign can act only when the populace is assembled." See *The Social Contract* (Rousseau 1987 [1762], p.71).

25. "Since the law is merely the declaration of the general will, it is clear that the people cannot be represented in the legislative power. But it can and should be represented in the executive power, which is merely force applied to the law" (ibid., p.75). Rousseau talks about elective aristocracy on p.54.

26. Rousseau (1987 [1762]), p.46. In Macpherson's view, Rousseau's *de jure* state would qualify as "one-class society." In Levine's view, on the other hand, Rousseau's vision is "petit-bourgeois" (Levine 1976, p.197). In Levine's view, Rousseau's vision, despite its "anti-capitalist" posture, does not take class divisions into account, and thus should be regarded as utopian (Levine 1976, p.187, also p.189, and p.197).

27. Rousseau (1987 [1762]), p.46.

28. Rousseau's sexism prevented him from including women in this definition of citizenship. In *Emile*, Rousseau expressed the view that "woman is specially made for man's delight," or she "is made to please and to be in subjection to man" (Rousseau 1995, Book V, p.385). In a book titled *Rousseau's Republican Romance* (2000), Elizabeth Rose Wingrove has argued that Rousseau's sexism is not just part of his personal character or beliefs, but more importantly, it forms the foundation of his political theory.

29. "As to the associates, they collectively take the name *people*; individually they are called *citizens*, insofar as participants in the sovereign authority, and *subjects*, insofar as they are subjected to the laws of the state" (Rousseau 1987 [1762], p.25, original italics.)

30. For Rousseau, democracy consists in "the sovereign entrust[ing] the government [i.e., the execution of the laws] to the entire people or to the majority of the people" (ibid., p.54). In this sense of the term, he believed that "true democracy has never existed and never will" (ibid., p.56). He also believed that in its true sense democracy was not practical: "It is unimaginable that the people would remain constantly assembled to handle public affairs" (ibid.).

31. Ibid., p.84.

32. Ibid., p.54.

33. Ibid.

34. Held (1987), p.74.

35. What follows is a statement about the absolute power of the sovereign. "Just as nature gives each man an absolute power over all his [bodily] members, the social compact gives the body politic an *absolute power* over all its members, and it is the same power which, as I have said, is directed by the *general will* and bears the name *sovereignty*" (Rousseau 1987 [1762], p.32, italics added).

36. These passages are quoted from *The Social Contract* (1762), p.26, p.27, and p.31, respectively.

37. A sympathetic interpretation of the statement that "the general will is always right" would be to say that a decision made by a large group of well educated and free people—using a proper decision-making scheme—cannot be wrong. If a majority-rule scheme is used to reveal the general will, then the proposition can be revised to state that the "majority is right." According to Levine, Rousseau would have agreed to this proposition, provided that the majority is "properly interrogated" (Levine 1976, p.68). Another plausible interpretation of this claim is Weale's example of a person who wears a suit to a party and once he gets there, he finds everyone dressed in jeans. This is an indication that he was mistaken (Weale 1999, p.26). In other words, the general will is right in the sense that given any issue and a binary choice, one choice would represent the common good better than the other or would be the only right choice. Being right in this sense would mean having the ability to see the better choice, or the right choice. According to Rousseau himself, in voting, wherefrom "the declaration of the general will is drawn from the counting of votes," when "the opinion contrary to mine prevails, this proves merely that I was in error, and that what I took to be the general will was not so" (Rousseau 1987 [1762], p.82).

38. While acknowledging that Rousseau must be regarded as a democrat, Melzer argues that he should also be characterized as a "defector" from the liberal camp (Melzer 1990, pp.109-110).

39. Rousseau (1987 [1762]), p.46.

40. Ibid.

41. Ibid., p.27.

42. Ibid., pp.26-27.

43. Ibid., p.27.

44. Held (1987), p.79.

45. As Rousseau understands it, when in the act of revealing the general will, i.e., in voting, if my opinion does not prevail, this simply means that "I was in error . . . if my private opinion had prevailed, I would have done something other than what I had [really] wanted. In that case, I would not have been free" (Rousseau 1987 [1762], p.82). It sounds as if Rousseau is saying that we should be thankful for being proven wrong, so we can want what we *really* want and be free.

46. In Chapter 7 of Book II in *The Social Contract.* Rousseau discusses the *de jure* state's responsibilities insofar as achieving the preponderance of the general will over private wills through discussing what the founding father of the *de jure* state is to achieve:

> He who dares to undertake the establishment of a people should feel that he is . . . in a position to change human nature, to transform each individual (who by himself is a perfect and solitary whole), into a part of a larger whole from which this individual receives . . . his life and his being; to alter man's constitution in order to strengthen it; . . . he must deny man his own forces [private wills] in order to give him forces that are alien to him and that he cannot make use of without the help of others. The more these natural forces [private wills] are dead and obliterated, and the greater and more durable are the acquired forces [general will], the more too is the institution solid and perfect (Rousseau 1987 [1762], p.39).

47. "Charismatic" is a characterization used by Macpherson (Macpherson 1965, p.19). The remaining characterizations are Rousseau's own (Rousseau 1987 [1762], p.38-39).

48. For Rousseau, 'the legislator' is "a superior intelligence that beheld all the passions of men without feeling any of them; who had no affinity with our nature, yet knew it through and through; whose happiness was independent of us, yet who nevertheless was willing to concern itself with ours; finally, who, in the passage of time, procures for himself a distant glory, being able to labor in one age and find enjoyment in another. God would be needed to give men laws" (Rousseau 1987 [1762], pp.38-39).

49. Ibid., p.39.

50. Ibid., pp.39-40.

51. Ibid., p.40.

52. Germino (1979), p.194

53. Ibid.

54. Rousseau's discussion of civil religion in *The Social Contract* takes place in Book IV, Chapter VIII. In discussing the relations between politics and religion earlier in the book, Rousseau argues that "one can serve as an instrument of the other" (Rousseau 1987[1762], p.41).

55. Levine (1976), p.187. The reason the legislator resorts to religious deception is that the intellectual deficit of the people prevents them from grasping the wisdom of the laws he proposes. In order to be heard and complied with, the legislator gives religious appeal to his edicts. The legislator must put his decisions, in the words of Rousseau himself, "in the mouth of the immortals, in order to compel by divine authority those whom human prudence could not move" (Rousseau 1987[1762], p.41). In a footnote, Rousseau quotes Machiavelli to the effect that "there has never been among a people a single legislator who, in proposing extraordinary laws, did not have recourse to God, for otherwise they would not be accepted" (ibid.).

56. Pateman (1970), p.26, original italics.

57. Ibid., p.26.

58. Ibid., p.27.

59. Ibid., p.27.

60. Ibid.

61. Rousseau (1987 [1762]), p.32.

62. "If, when a sufficiently informed populace deliberates, the citizens were to have no communication among themselves" (ibid.)

63. Ibid.

64. Ibid., p.38.

65. Ibid., p.31.

66. Ibid., p.38.

67. Femia (1996), p.389.

68. Ibid. Femia takes the quotation from Rousseau's *A Discourse on Political Economy*.

69. Germino (1979), pp.193-94.

70. Neidleman (2001), p.66. Also see p.2.

71. Ibid. p.2.

72. Ibid. See also p.5.

73. One should add that the idea behind the term "reasonable" utilized here is similar to the Rawlsian idea of reasonable as Rawls himself applies it to society and persons. That is to say, a popular will could be regarded as reasonable if it represents the desire for a fair system of social cooperation. A will can be said to be reasonable if the citizens who

form it, to use Rawls' description of "reasonable persons," "desire for its own sake a social world in which they, as free and equal, can cooperate with others on terms all can accept" (Rawls 1993, p.50). As reasonable persons, those who form a reasonable will, would be capable of being persuaded to "discuss the fair terms that others propose" because they, as knowledgeable individuals, understand the importance of their sociality, as indicated by their desire for social cooperation (ibid., p.49).

74. Needless to say, Rousseau did not picture this image for his *de jure* state, for he did not allow the assumptions of politically knowledgeable and cognitively competent citizens; nor did he permit public deliberations. There are ample reasons to indicate that Rousseau truly believed that the ordinary citizens lacked sufficient intelligence. Trachtenberg argues that Rousseau was of the view that ordinary citizens suffered from "cognitive deficit" and "the legislator" was his solution to this problem, and that this solution is in line with Rousseau's "intellectual elitism" which he had established in his *First Discourse* (Trachtenberg 1993, p.239). Putting Rousseau's problem of seeking a synthesis of popular will and rational will within the context of the fundamental political problem of "consent versus wisdom," Melzer argues that in Rousseau's own solution to the problem, "the public title of wisdom [i.e., rational will] is completely rejected and all official respect and legitimacy is given exclusively to the principle of consent [i.e., popular will]; but at the same time, wisdom, which is indispensable to the state, is quietly urged to exercise a vast indirect power, through the [legislator's] covert manipulation of the single official authority, the popular will" (Melzer 1990, p.241).

75. Rousseau's banning of public-dialogical forms of political education for adult citizens seems to parallel the curriculum he recommended for the public education of children. In his examination of Rousseau's *Considerations on the Government of Poland*, Trachtenberg comments that Rousseau "places very little emphasis on developing children's cognitive abilities" and that he sees public education consisting in "the absorption of facts [about one's country] rather than the development of intellectual abilities" (Trachtenberg 1993, p.234). For Rousseau, "cognitive development is not the goal of public education at all" (ibid., p.237). "Rather," Trachtenberg continues, "the function of public education is to instill civic virtue, in particular the disposition to obey the law. . . . Students are discouraged from thinking (even day-dreaming) for themselves, and instead are taught to base their actions on others' opinion." (ibid.) Trachtenberg concludes that Rousseau's view on public education "does not address the issue of *formulating* the general will. It does not provide training in the skills required to discover the common good—but only in the attitudes that motivate individuals to seek the common good once it has been pointed out to them" (ibid., p.238, original italics).

76. In any democratic reinterpretation of Rousseau's conception, as in the one attempted above, the legislator would need to share the responsibility of educating citizens on civic and political matters with some other (independent) individuals who also possess high moral and intellectual competencies. Since the legislator's main *raison d'être* is to offer political wisdom and moral guidance to the public in getting to know the general will and thus ensuring that the laws made are "good" laws, there is no reason to believe that this function could be served best by a single individual. Thus, there might as well be more than one grand legislator. Taking the latter as the case, though each legislator might differ from others on what the general will ought to be in each voting situation, in educating citizens on issues, each and all would proceed with the commitment that they ought to direct citizens to honor their civic responsibilities, and in voting on issues, to place their conceptions of generalizable interests above their private interests. Furthermore, there are no good reasons to believe that the legislator (or legislators) could perform his (their)

educational functions best by intimating their wisdom and guidance rather than by communicating it dialogically by the force of rational arguments. As was argued in note 74 above, Rousseau's "intellectual elitism" and his view that ordinary citizens suffered from "cognitive deficit" would have prevented Rousseau from being persuaded by the latter argument. In Rousseau's view, this deficit of citizens would prevent them from grasping the wisdom, and appreciating the higher reason, behind what the legislator would intimate to them as the general will. Melzer has argued that Rousseau believed "if the latter [the legislator] must rely solely on rational argument, he will find it quite impossible to convince the people to consent to his salutary but necessarily strange, austere, and demanding system of laws" (Melzer 1990, p.235). And this is why, in order to convince the people of the righteousness of what he intimates to them as the general will, the legislator has to resort to "fraud, to win the people's consent through religious deceit" (ibid.). The reason Rousseau did not favor the rational-cognitive development of citizens, and instead, exclusively emphasized their civic-moral development was that he believed such a development would direct citizens to focus on their individuality and self-interested pursuits, and this was exactly what the Enlightenment had done. As Allan Bloom contends, Rousseau objected to the Enlightenment because he believed it dissolved "habits of sentiments" that were "vulnerable to reason, which sees clearly only calculations of private interests" (Bloom 1997, p.148).

Chapter 4

The "Marxian Idea of Democracy": The Ideal and the Real

In continuing to explore Macpherson's notion of democracy as a "class affair," one now needs to turn to Marx and Engels. The "Marxian idea of democracy," according to Macpherson, "started from the age-old notion of democracy as class rule but gave it a new turn by making it more precise."[1] Macpherson quotes a famous passage from *The Manifesto of the Communist Party* that states: "the first step in the revolution by the working class, is to raise the proletariat to the position of the ruling class, *to win the battle of democracy*."[2] Beyond this, Macpherson does not say much about the Marxian idea of democracy and instead begins to examine "the communist conception of democracy" in the Soviet Union. This chapter is devoted to examining the Marxian idea of democracy. (A discussion of the Soviet concept will be presented in the ensuing chapter.)

Returning to the passage quoted above from the *Manifesto*, despite this reference to democracy in a crucial passage in their most important work, it is a well-known fact that Marx and Engels made no attempts at developing a theory or a concept of democracy. Marx and Engels were not particularly interested in discussing the question; nor did they make any pretensions to being democrats. Whenever they stumbled upon the question of democracy, or were forced to address it, they, speaking in general terms, either mocked and sneered at those who were seriously devoted to the idea, or subjected the idea to harsh criticisms. This is mainly because Marx and Engels took the word democracy, in general, to mean "bourgeois democracy"—which they believed was an unworthy cause to struggle for—and rejected all such struggles as "illusory" and naïve.

In order to develop an understanding of why Marx and Engels dealt with the question of democracy in this way or what they took democracy to mean to begin with, perhaps the best way to proceed would be to catalog their rare and scattered uses of the terms democracy and democrats into two periods; first being their younger years before they embraced communism, and the other, their more mature and communist period.

There are reasons to believe that in the first period, Marx had a positive and idealistic notion of what democracy meant, or ought to mean, and believed that democracy was an ideal that society ought to realize. In a letter written to Arnold Ruge in May 1843, Marx gives the impression that he takes democracy or a "democratic state" to be a "community of people" or of "free men" who have a "sense of self-worth," and work toward the realization of their "highest ends"—Marx also implies here that "the Greeks" had such a community.[3] In the same letter, Marx's usage of the term democracy also conveys the impression that he believes humanity, in its most advanced form, would cross over from "the political animal world" into "the human world of democracy."[4] (One cannot help but see the parallel between this statement and what he wrote in 1859 in the "Preface to the Critique of Political Economy," to the effect that crossing over to communism would bring "the prehistory of human society to a close."[5]) A month or two later, in his *Critique of Hegel's 'Philosophy of Right'*, he presented in an abstract theoretical language another positive and idealistic understanding of what he takes democracy to be. There, in comparing democracy with other political systems, Marx expresses the view that "[d]emocracy is *human existence*, while in the other political forms man has only *legal* existence. That is the fundamental difference of democracy."[6] Moreover, unlike monarchy which is, or should be, only a form, "[d]emocracy is [both] content and form."[7] He also comments that "in democracy the constitution itself appears only as one determination, and indeed as the self-determination of the people."[8] Furthermore, "[d]emocracy starts with man and makes the state objectified man"—here Marx is attempting to invert Hegel's political philosophy who "proceeds from the state and makes man into the subjectified state."[9] Unclear as they may be, it seems that these are the only places in the entire writings of Marx (both young and mature)—and perhaps Engels as well—that one finds an attempt to offer an explicit (and positive) expression of what he took democracy to represent as an ideal.[10]

In the second period, starting with *The German Ideology* in the summer of 1846, one encounters a different usage of the term democracy in the works of Marx and Engels. In this period, again, in the absence of explicit definitions or expressions, and based on what is available, one can only infer that they held the view that the genuine interest in the struggle for democracy arises from a misguided, "illusory," and naïve understanding of the social world, as well as from a blindness to the fact of class divisions in society. Attribution of this view to Marx and Engels is justified on the basis of their characterization of those who were struggling for democracy, "the democrats," as "blind," "vulgar," and "light-minded" people who harbor "illusions" and "delude" themselves and fall for "ideological nonsense about rights and other trash" of the capitalist class. In addition to these characterizations, sometimes Marx and Engels humor the democrats as "honest men" and philanthropists.[11] As to regarding democracy as an illusory form of struggle, one such evidence can be found in *The German Ideology* where Marx and Engels comment that "all struggles within the state, the struggle between democracy, aristocracy, and monarchy, the struggle for the franchisee . . . are merely *illusory* forms . . . in which the real struggles of the different classes are fought out among one another."[12]

Another piece of evidence to support this view can be found in *The Eighteenth Brumaire of Louis Bonaparte*, where Marx rebukes and mocks "the democrats" for turning a blind eye to the fact that their enemy is a "privileged class," and "they, along with all the rest of the nation, form the *people*. What they represent is the *people's rights*; what interests them is the *people's interests*. Accordingly, when a struggle is impending, they do not need to examine the interests and positions of the different classes."[13] Marx comments earlier in the same passage that "[n]o party exaggerates its means more than the democratic, none *deludes* itself more *light-mindedly* over the situation."[14] Moreover, in the introduction to the 1895 edition of *The Class Struggle in France,* in summing up the lessons learned from the defeats suffered in the struggles of 1849, Engels for his part takes note of the *"illusions"* of "vulgar democrats" who "reckoned on a speedy and finally decisive victory of the *'people'* over the 'tyrant'; [while] we looked to a long struggle, after the removal of the 'tyrant' among the antagonistic elements concealed within this 'people' itself."[15] Furthermore, in a famous passage in the *Critique of the Gotha Program*, Marx expresses the view that "the democrats" are inclined to fall for "ideological nonsense about rights and other trash."[16] Finally, in a joint letter with Engels, Marx seems to mock those "honest men" who are for "bourgeois democracy," for it is a philanthropic movement.[17]

In contrast to this naiveté of "the democrats," Marx and Engels seem to have believed that the struggle for democracy, as they put forth in the *Communist Manifesto* in 1848, ought to *really* mean a struggle "to raise the proletariat to the position of the ruling class."[18] This is *the way* "to win the battle of democracy."[19] However, this assertion does not say much about how democracy would relate to the question of the class rule of the proletariat or whether the proletariat in the position of power would rule "democratically." In order to develop an understanding of how they would have possibly addressed these questions, one needs to turn to Engels' *Principles of Communism,* written almost two months earlier. According to Engels,

> In the first place it [the proletariat revolution] will inaugurate a *democratic constitution* and thereby directly or indirectly the political rule of the proletariat.[20]

Engels continued.

> Democracy would be quite useless to the proletariat if it were not immediately utilised as a means of accomplishing further measures *directly attacking private ownership* and *securing the existence of the proletariat.*[21]

What exactly did Engels mean by "democratic constitutions"? What did he think this democratic constitution would contain? Will there be any guarantee of equal political rights and freedoms for all citizens? How will the proletariat in power rule "democratically"? Engels did not specify. However, given the fact that the passage quoted above is followed by a twelve-point program of socialist measures that the proletariat must carry out—and none of them have anything to do with the protection of equal political rights and freedoms for all citizens—one is inclined to think that by "democratic constitution," Engels mainly meant a constitution that gives sweeping rights and freedoms to the state to take socialist measures. This evaluation is bolstered in view of what he says a few pages later, where he characterizes the proletariat's rise to power as the *"winning of democracy"* and implies that this necessarily should be followed by socialist measures: "the realisation of socialist measures which must follow it."[22] However, a few pages further down, in discussing the path to the proletariat victory in the United States, Engels evaluates the American constitution as a "democratic constitution," but quickly adds that "the Communists must make common cause with the party that will apply this constitution against the bourgeoisie and use it in the interests of the proletariat, that is, with the national agrarian reformers."[23] One should bear in mind that Engels here has the pre-Civil-War United States as his context, where slavery was in full swing. Thus, Engels' notion that the U.S. Constitution should be applied toward the interests of the proletariat should be understood primarily as advocating the abolition of slavery and supporting the land reform that might follow it. The evaluation suggested above—that Engels' notion of the democratic constitution is not about equal political rights and the freedom of all citizens—is further strengthened by his way of characterizing the use of the state by the proletariat in power. According to Engels, "the proletariat uses the state not in the interests of freedom but in order to hold down its adversaries."[24] In light of these considerations, one is justified in concluding that the democratic constitution Engels has in mind would grant sweeping rights to the political rulers of the new state to use coercive power to subordinate the capitalist class and enable the new regime to defend itself against potential bourgeois counter-revolutions, and thus secure its existence. As will be shown later, it was this anti-bourgeois and "securing" aspect of Engels' thought on the question that would later prevail in Lenin's understanding of democracy.

What is often regarded as a model for the Marxian understanding of how the proletariat would rule once it is placed in the position of power—or what a "democratic constitution" should look like, or what should constitute a *real* democracy in action—is Marx's commentary on the Paris Commune in *The Civil War in France,* where he spoke of the Commune in glowing terms and saw it as trumpeting the rise of a new society on the horizon. (Engels later referred to the Paris Commune as "a new and truly democratic [state]."[25]) "The Commune," as Marx understood it, "was to be a working, not a parliamentary, body, executive and legislative at the same time."[26] All elected officials were to be revocable.[27] The Commune was, to use David Held's characterization, a "direct democracy."[28] It gave the workers "[f]ull local autonomy within [a] framework of councils" that had a "pyramid structure."[29] In short, the Commune can be characterized *as a sys-*

tem of rule by the elected officials who were under the direct control from below
(the majority of the population, viz., the working-class people and their allies).

One way to reconcile the young Marx's *ideal* and positive notion of democracy with the mature Marx and Engels' sense of what it *really* ought to mean is to argue that *Marx never gave up the ideal notion, but believed that it could not be achieved without first abolishing classes.* For Marx, the *real* content of democracy was its class content and not its "procedural" form that blind, vulgar, and light-minded democrats took for its content. Pushing this view further, one can attribute to Marx the view that democracy was only realizable in the higher phase of the communist society.[30] One version of this view is expressed by Maximilian Rubel. According to Rubel in his "Marx's Concept of Democracy," "[f]ar from breaking with his first conception of democracy [in the first period] when he became a communist, Marx sublimated it. In communism as he understood it, democracy was not only maintained, but acquired even greater significance."[31] Taking for granted that Marx believed the communist society was a democratic society, David Held argues that Marx was completely sold on the idea of *direct* democracy as it was practiced, although for a short period, in the Paris Commune in 1871. According to David Held, a "direct democracy" of the sort practiced in the Commune was what he had in mind "[w]hen Marx referred to 'the abolition of the state' and the 'dictatorship of the proletariat'" in his writings in the post-Commune period.[32] On this, Held is probably following Engels who concluded his introduction to the new edition of *The Civil War in France* on March 18, 1891, with the following statement:

> Dictatorship of the proletariat. Well and good, gentlemen, do you want
> to know what this dictatorship looks like? Look at the Paris Commune.
> That was the dictatorship of the proletariat.[33]

However, the view that Marx could have believed that the social organizational form of the Paris Commune was to be held as a model for *both* phases of the communist society seems to be questionable. What seems to be a more acceptable view on this matter is to say that most likely, Marx believed that, as Held argues, "the Commune provides a definite model for at least the 'first stage of communism'."[34] Held is also of the view that both Marx and Engels took the Athenian model of democracy as "a source of inspiration . . . [and] . . . their own model of properly democratic order, the Paris Commune, . . . suggests a remarkable number of common features with Athens."[35] One should keep in mind that the most important common feature of these two models was the integration of the legislative and executive functions.

To bring this chapter to a close, the most plausible and tenable position to take on the question of Marx and Engels' view of democracy would be to say that they equated *real* or "true" democracy with the dictatorship of the proletariat. In this sense, democracy in the post-revolutionary society would start as a class rule—and with a new "democratic constitution" that would function as an apparatus to institute the abolition of private property and secure the survival of the proletariat's state—and would work its way toward establishing a classless society. Once this state of affairs is achieved, i.e., in the highest phase of the communist society, the state in its traditional meaning of class rule will disappear, and the "administration of things" by citizens, as Engels once put it, would begin. However, it should be stated that at this point, there seems to exist a point of controversy over whether to characterize the communist society as truly democratic—(in the case of the young Marx's ideal, as his letter to Arnold Rouge seems to be hinting at, or as Rubel has argued)—or as the state of the abolition of democracy as Engels seemed to hint in his preface to *The Pamphlet Internationales Aus Dem "Volksstaat"* on January 3, 1894. In this preface, after mentioning that he calls himself a Communist instead of a Social Democrat, Engels goes on to say that the "ultimate political aim" of the Communist Party "is to surpass the entire State, and thus democracy too."[36] Given the fact that surpassing of the state means its abolition, it is possible to argue, as Lenin did almost two decades later, that "the abolition of the state means also the abolition of democracy; that the withering away of the state [also] means the withering away of democracy."[37]

Notes

1. Macpherson (1965), p.15.

2. Ibid., italics added.

3. Padover (1979), p.25. The full quotation is as follows. "Men—means intellectual beings; free men—means republicans. . . . One will have to reawaken in the breast of these people ["the common philistines"] the sense of self-worth of men—freedom. Only such sense, which vanished from the world with the Greeks and evaporated into the blue with Christianity, can transform society again into a community of people for their highest ends—a democratic state."

4. Padover (1979), p.27.

5. Marx and Engels (1977), Vol.1, p.504.

6. Marx (1970), p.30, original italics.

7. Ibid., p.28.

8. Ibid., p.29.

9. Ibid., p.30.

10. Douglas Lummis (1995) regards Marx as a "radical" democrat based on his reading of the *Critique of Hegel's 'Philosophy of Right'* and comments that the writing is "[t]he closest work . . . to a manifesto for radical democracy" (ibid., p.167).

11. It should be added that by "the democrats" in their various writings, Marx and Engels meant different groups of people. But what was common to all of them, as Marx and Engels saw them, was that they were not involved in a class-struggle against the established political power.

12. Marx and Engels (1976), p.52, italic added.

13. Marx and Engels (1977), Vol.1, p.427, original italics.

14. Ibid., p.426, italics added

15. Ibid., p.189, italics added.

16. Marx and Engels (1977), Vol.3, p.10. The actual sentence: "ideological nonsense about rights and other trash so common among the democrats and French Socialists."

17. Ibid., p.89.

18. Marx and Engels (1977), Vol.1, p.126.

19. Ibid.

20. Ibid., p.90, original italics.

21. Ibid., italics added.

22. Ibid., pp.95-96.

23. Ibid., p.96.

24. In a letter written to August Bebel, Engels goes on to say that "and as soon as it becomes possible to speak of freedom the state as such ceases to exist." This quotation is cited by Lenin in the *State and Revolution* (Lenin 1952, Vol.II, Part 1, p.291). Lenin does not provide the date of the letter.

25. Marx and Engels (1977), Vol.2, p.188, in the introduction to the new edition of *The Civil War in France*, dated March 18,1891. The full quotation: "This shattering [*Sprengng*] of the former state power and its replacement by a new and truly democratic one is described in the third section of *The Civil War.*"

26. Ibid., p.220.

27. Ibid.

28. Held (1987), p.130.

29. Ibid., pp.126-27. (See the first sentence of the first paragraph on p.130.)

30. Allen Wood comments that Marx's disinterest in discussing the question of "democratic" control of society in the postcapitalist society should be attributed to the fact that "he does not see the problem as a procedural one at all" and that Marx saw "the chief obstacle both to individual freedom and social unity . . . [in] . . . the division of society into oppressing and oppressed classes" (Wood 1984, pp.57-58). This can be taken to mean that Marx was not interested in the question of the "form" of democracy, as "vulgar democrats" did, as much as he was interested in its class "content."

31. Rubel (1983), p.103. Rubel believes that Marx's ideas of socialism and communism "took shape" in his thought "within the notion of democracy" (ibid.).

32. Held (1987), p.128. Held argues that "Engels was certainly of this view" (ibid., p.128n). Moreover, Held characterizes Marx's model of democracy as "direct democracy" (ibid., p.130).

33. Marx and Engels (1977), Vol.2, p.189. In the preceding paragraph of the same writing (p.188), as was quoted earlier, Engels had characterized the Commune as "a new and truly democratic [state]."

34. Held (1987), p.128n.

35. Held (1987), p.21. It seems that Held is basing this claim primarily on his examination of Marx's letter to Arnold Ruge in May 1843 that was considered earlier. Held quotes Marx in 1842 to the effect that Marx was nostalgic about the "self-reliance" and "freedom" of the Greeks (ibid., p.130). However, it seems that Held is mistaken and the quotation comes from the above-mentioned letter. For Held, the combining of the legislative and executive functions of the state is one of these common features (ibid., p.128). (See Held's long quotation from *The Civil War in France.*) Finally, Held is quick to point out that the model Marx had suggested on the basis of the experience of the Paris Commune, would have been a "highly *indirect* form of democracy" if it were to be expanded throughout the nation (ibid., p.131n, original italics).

36. Marx and Engels' *Collected Works* (1975), Vol.27, p.417.

37. Lenin in the *State and Revolution* (Lenin 1952, Vol.II, Part 1, p.284).

Chapter 5

Bolshevism and the Idea of "Proletarian Democracy"

The term Bolshevism here is taken to mean in a broad sense the theory—and to some extent the practice—of the October Revolution of 1917, from its beginning days until its betrayal by Stalin's consolidation of power in the late 1920s. The central and dominant figure of Bolshevism, no doubt, was no one other than Vladimir Ilych Lenin. Bolshevism encompassed a wide range of social-political theories and revolutionary political principles on the questions of the state, society, and the vanguard party, as well as encompassing political strategies for leading to victory the revolution of the oppressed people in the most backward of the major countries in Europe. Treating Bolshevism also as encompassing a theory of democracy follows in the tracks of Macpherson's discussions regarding the "communist concept of democracy" in his *Real World of Democracy*.[1]

Based on the failure of the western European working-classes to rise to the position of the ruling class, Lenin concluded that, as Macpherson puts it, "the proletarian revolution would have to be the work of what he called a vanguard, a fully class-conscious minority."[2] In Lenin's strategy, the vanguard party, consisting primarily of revolutionary Marxists, would seize the state power with the support of the oppressed classes of the workers and peasants—who constituted the overwhelming majority of the Russian population. The revolutionary vanguard party would seize the state power and rule not only in the name of the oppressed people, but also ultimately in the name of entire humanity. The oppressed working-class of Russia at the time, as well as the peasantry, was not only incapable of developing its ascribed and potential class-consciousness—(the experience of the western European working-class had demonstrated that this was impossible in the age of imperialism)—but was also culturally and politically illiterate and backward, as it was economically downtrodden and impoverished. This latter problem, as Lenin and Trotsky got to appreciate its fullest scope only in the post-revolutionary period, compounded the difficulty of establishing a political system of a new type in

the midst of overwhelming various other problems, such as a civil war and a ru-
ined economy.[3] Given these circumstances, the cards were stacked against the
Bolsheviks from the very beginning, once they had managed to take the state
power in their hands.

> Instead of being able to start as a class democracy it [the new Soviet
> state] had to start as a vanguard state. It had to try to work towards a
> high-productivity classless society while it was making up the distance
> between the vanguard state and that full proletarian democracy which
> Marx had envisaged as the first stage immediately after the revolution.[4]

As it turned out, the new regime failed to travel the road from the vanguard
state to the "full proletarian democracy" as it was seized by counter-revolution in a
matter of a few years. The regime that ended up consolidating its power in the
territories that later came to be known as the Soviet Union, and ruled in the name
of the proletariat until its implosion in 1991, failed to qualify as a democratic state.
In the first place, it failed because, as will be argued later, it committed the ultimate Rousseauean sin of alienating and representing the sovereignty of the people
in general, and that of the working-class in particular. This failure of the new Soviet regime looked particularly egregious when it was beheld in the mirror of liberal democracy that co-existed with it in the West. Despite its betrayal, the significance of the October Revolution lies in that, for the first time in human history, a
vanguard revolutionary party, composed primarily of revolutionary intellectuals,
attempted to build a new society based on the idea of the substantive equality of
all, as well as on the basis of a complete devotion to the idea that "the free development of each is the condition for the free development of all."[5] Moreover, as
will be discussed in this chapter, the Soviet regime also failed the test of democracy in the sense of the view attributed to Marx and Engels earlier, and in the sense
of the "proletarian democracy" developed by Lenin himself.[6]

*   *   *   *   *

In light of the fact that Marx and Engels' writings lacked a coherent notion of what
they took democracy to be, the task of developing a Marxist position on the question of democracy fell on the shoulders of Lenin. Lenin's first serious attempt at
developing a Marxist viewpoint on the question can be found in his *State and
Revolution*, written in August and September of 1917, just a month or two before
the October Revolution. There, with Marx's *Critique of Gotha Program* in mind,
Lenin argued that,

> [d]emocracy means equality. The great significance of the proletariat's
> struggle for equality and of equality as a slogan will be clear if we cor-
> rectly interpret it as meaning the abolition of *classes*. But democracy
> means only *formal* equality. And as soon as [this] equality is achieved
> for all members of society *in relation* to ownership of the means of
> production, that is, equality of labor and wages, humanity will inevita-
> bly be confronted with the question of advancing further, from formal

equality to actual equality, i.e., to the operation of the rule "from each according to his ability, to each according to his needs."..[7]

Here "quantity turns into quality": *such* a degree of democracy implies overstepping the boundaries of bourgeois society and beginning its socialist reorganization. If really *all take part in the administration of the state*, capitalism cannot retain its hold.[8]

In these passages, Lenin, following in the footsteps of Marx and Engels, argues that the *real* content of the concept of democracy for the proletariat should be nothing other than the complete abolition of classes. Once the goal of the abolition of classes is achieved, that is to say, once the highest stage of communism is arrived at, the limiting and formal boundaries imposed on equality and democracy in their bourgeois understanding will be stepped over. At this stage, the meaning of democracy will go through a qualitative change to mean the participation of *all* in "the administration of the state." Moreover, following Marx in *The Civil War in France*, Lenin emphasizes that in the transitional stage, beginning *immediately* after the seizure of power, all of the state officials would be elected to their offices. Moreover, they would be subject to recall at any time, and their salaries would be at the levels of ordinary workers.[9]

Lenin also follows Marx in *The Civil War in France* on the following points.

Representative institutions remain, but there is *no* parliamentarism here as a special system, as the division of labor between the legislative and executive, as privileged position for the deputies. We cannot imagine democracy, even proletarian democracy, without representative institutions, but we can and *must* imagine democracy without parliamentarism.[10]

There can be no thought of abolishing the bureaucracy at once, everywhere and completely. That is utopia. But to *smash* the old bureaucratic machine at once and to begin immediately or to construct a new one that will permit to abolish gradually all bureaucracy—this is *not* utopia, this is the experience of the Commune, this is the direct and immediate task of the revolutionary proletariat.[11]

Obviously, Lenin believed that the *participatory* democracy in communism, as he envisioned it (i.e., "*all* take part in the administration of the state") was a good thing, and perhaps one could assume that Lenin took this as the *ideal* meaning of democracy. In order to gain some insight into Lenin's understanding of *real* or "proletarian democracy" in the immediate post-revolutionary period, one needs to turn to his contribution to an international debate on the question of the dictatorship of the proletariat that took place between him and Karl Kautsky in late 1918.[12] The full presentation of the views expressed by Lenin in this debate is beyond the scope of the task at hand. Here, it suffices to mention that Lenin used the occasion to pound in his pamphlet (*Proletarian Revolution and the Renegade Kautsky*) the point that democracy has always been and will be a "*class* democracy."[13] Moreover, he charged that "[b]ourgeois democracy . . . cannot but remain . . . restricted, truncated, false and hypocritical, a paradise for the rich and a snare and a deception for the exploited, for the poor."[14] Lenin also led an attack on the "formal equality" of bourgeois democracy as he argued that,

> [e]ven in the most democratic bourgeois state the oppressed masses at every step encounter the crying contradiction between the *formal* equality proclaimed by the "democracy" of the capitalists and the thousands of *real* limitations and subterfuges which turn the proletarians into *wage slaves.*[15]

The true merit of the Soviet system, Lenin argued, lies in that it gives a real meaning and content to the "right of assembly" for the workers.[16] In defending the absence of direct elections for "nonlocal Soviets," Lenin argued that indirect elections "make it easier to hold Congresses of Soviets, they make the *entire* apparatus less costly, more flexible, more accessible to the workers and peasants . . . to be able very quickly to recall one's local deputy."[17] Thus, the Soviet system provides *"far more accessible* representation" to the people (workers and peasants).[18] These contentions led Lenin to proclaim that "[p]roletarian democracy is *a million times* more democratic than any bourgeois democracy."[19] This is because the proletarian democracy is a *"a democracy for the poor."*[20] (One should note that Lenin said a democracy *"for"* the poor and not *"by"* the poor.) And if Kautsky does not understand this, it is because "[h]e fails to see the *class* nature of the state apparatus, of the machinery of [the Soviet] state."[21]

In his attacks on Kautsky in *The Proletarian Revolution and the Renegade Kautsky*, Lenin's argument is persuasive: to exercise proletarian democracy as he defines it, one cannot and should not grant political equality to all citizens. This means that the oppressors must lose their political rights in order to permit the workers' state to go ahead and accomplish, using Engels' words quoted earlier, "further measures directly attacking private ownership and securing the existence of the proletariat." On this point, Kautsky's bourgeois democratic point of attack thus can be deflected successfully. As Lenin emphasized, the democratic backbone of the regime he was presiding over was the network of the Soviets. On the strength of the Soviets—to the extent that workers and peasants had a real voice in the Soviets and that the Soviets could exercise real influence and/or control over the institutions that exercised the dictatorship of the proletariat (the highest one being the party)—Lenin was well justified in declaring the Soviet system as democratic, as well as being more democratic than the Western model, where the interests of the bourgeoisie were voiced and those of the workers were trampled upon. Moreover, in light of what was presented, Lenin, in the works cited, did not diverge from the teachings of Marx and Engels in theory. What was new in Lenin was that he placed the vanguard party of the professional revolutionary Marxists in the position of guiding the government of the Soviets.

Having presented this overview of Lenin's conception of the proletarian democracy, it should be added that some socialist scholars (e.g., Samuel Farber, Ralph Miliband, and Carmen Sirianni) have argued, and rightly so, that Lenin's conception of what constitutes a real democracy (i.e., proletarian or Soviet democracy) in the pre-revolutionary period, as outlined earlier, was flawed. This is because Lenin's conception understood democracy mainly in terms of the class-based and "apolitical administration" of society, and at the same time lost sight of the *political* character of the concept.[22] Lenin's conception had no understanding of, in the words of Farber, "political processes to settle the inevitable differences of opinion within the working class," as well as procedures and "socialist legality."[23] To "conceive of democratization as appointing people with a working-class or peasant background to governmental and economic administrative positions," as Lenin did, is to commit a serious theoretical mistake.[24] As to the question of the "socialist legality" or the question of establishing theoretical-legal criteria or limits for the measures that the dictatorship of the proletariat, or the proletarian democracy for that matter, would be allowed to take, it should be pointed out that Lenin's long-standing position was that the dictatorship meant unlimited power based on force and "not on law."[25]

*   *   *   *   *

Putting aside the flaw in his conception of democracy in the pre-revolutionary period, what proved to be as fatal, if not more so, to Lenin's vision of proletarian democracy, outlined in the *State and Revolution*, were the practical problems that the new Soviet regime faced. The unfavorable objective conditions that the new regime found itself in—(in particular, a predominantly agrarian and under-productive economy in shambles at the end of World War I, the destruction of the Civil War, and invasion by foreign armies)—forced it to begin with emergency measures at its moment of inception. In moving from one emergency to another, improvised policies, one after another, took the place of developing policies on the basis of the principles that were adhered to, and views that were expressed, in the pre-revolutionary period. From the very beginning, the new regime had a siege mentality and was more concerned with its own survival than with organizing the Russian proletariat to be at the helm of exercising the dictatorship of the proletariat. The new regime soon began ruling by decrees. Already by the end of the first year in power, the dictatorship of the proletariat had been transferred into a "specially organized dictatorial regime."[26] Under the pressure of the unfavorable objective conditions, it soon became evident that the Bolshevism in power would fail to live up to the model of the Paris Commune regarded by Marx and Engels as a model for the dictatorship of the proletariat.[27] It was under the pressure of these circumstances that Lenin's theoretical inconsistencies and difficulties with the Marxist theory began.[28]

In addition to the unfavorable objective conditions that contributed to the distortion of principles, there was an unfavorable subjective factor that also proved fatal in practice—(perhaps as fatal as the objective factors)—to the goal of expanding proletariat democracy that by original design was intended to be exercised through the network of the Soviets. This factor was none other than the political-cultural backwardness of Russian society at the time. This cultural factor should definitely be related to the discussion of Lenin's understanding of what democracy should really have meant in Russia of the time. Moreover, the factor should also be considered in discussing the ultimate failure of the Bolshevik project of actualizing the idea of the proletarian democracy. The scope and depth of the cultural problem came to the forefront only after the victory of the revolution, and especially in light of the obstacles encountered in the course of the implementation and execution of War Communism during the Civil War. More than any other leading Bolshevik, Trotsky paid special attention to this problem. A cursory reading of his *Problems of Everyday Life* would reveal the wide scope of the cultural issues that Trotsky and other Bolshevik leaders had to address.[29] "When we created the state," as Trotsky stated on one occasion, "only then we realized properly for the first time how much we lag behind, how little culture we have. And the most elementary problems stood before us in all their concrete immensity."[30]

Lenin's appreciation of the depth of the poverty of culture in Russia—and the recognition of the need to address it—came about gradually, and peaked only when he had become, for all practical purposes, politically ineffective. This recognition prompted him to express the need for a "cultural revolution."[31] In his last writing, Lenin stated that "[w]e . . . lack enough civilization to enable us to pass straight on to socialism, although we have the political prerequisites for it [i.e., the state power]."[32] As was the case with Trotsky, Lenin too understood the question of culturalizing the Russian people—even when he has fully grasped the depth of the problem—mainly in terms of enriching their "material culture" and to a lesser degree, their "social culture."[33] The main motive behind the idea of culturalizing Russia was to turn Russian workers and peasants into productive citizens and willing participants in building a viable state-administered economy. Arguably, this level of involvement was the extent of Lenin's post-revolutionary vision of the people's participation in the new regime.

As to the question of *political culture* or culturalizing Russian people politically, Bolsheviks worked with what they had inherited from the old society. Thanks to a long tradition of political oppression and autocracy, pre-revolutionary Russia was virtually devoid of anything that could be properly called a "democratic" political culture or tradition. Habits of participating in political campaigns, voting in elections, tolerating the views of opponents, building consensus, adhering to procedures, and accepting electoral defeats were alien notions in the political culture that Bolsheviks had inherited. Against this backdrop, it would be fair to say that Lenin saw the question of the political culturalizing of the ordinary people largely along the lines of enabling them to stand up to the corrupt and autocratic culture of the Czarist bureaucracy, which was still in operation, and forcing it to reorganize itself into a small workers' state machinery that would serve the people (mostly peasants) and earn their confidence.[34] What is noticeably absent here is any trace of the idea of direct and participatory measures Lenin had expounded in the *State and Revolution* on the eve of the October Revolution.

By many indications, the cultural backwardness of Russia was a main reason for Lenin's hesitation to fully embrace the idea of direct (Soviet) democracy and expand it to the degree he had envisioned in the *State and Revolution* or in his response to Kautsky. It was not just the pressure of the emergency circumstances that led him to distort or revise the principles he held in the pre-1918 period, but also his diminished confidence in the Russian proletariat's political-cultural competence. Perhaps this is why he was moving in the direction of theoretically altering the meaning of the dictatorship of the proletariat to mean the dictatorship of only its "revolutionary elements."[35] In an address delivered in March 1919, Lenin argued that given the "low cultural level" of Russia,

> the Soviets, which by virtue of their program are organs of government by the working people, are in fact organs of government for the working people by *the advanced section of the proletariat*, but *not by the working people as a whole*.
>
> Here we are confronted by a problem which cannot be solved except by prolonged education.[36]

Needless to say, the "prolonged education" can only bear fruit in a distant future. In the meantime, "the advanced section," the party, had to take upon itself to exercise the dictatorship of the proletariat in the name of the whole class. As it turned out, Lenin's gravitation toward this line of thinking on the question of the dictatorship of the proletariat had unfortunate consequences for the Russian Revolution in the post-Civil War era. Lenin's tendency to resort to "administrative and police measures"—rather than using "political means"—in dealing with the opponents of the regime in this period, one would suspect, was a direct consequence of this line of thinking on the question of the dictatorship of the proletariat.[37] By 1922-23, according to Farber, the Soviets had lost a large portion of their powers to the party.[38] Moreover, the decline of the power of the Soviets in this period had paralleled a rapid decline of internal democracy inside the party.[39]

*   *   *   *   *

To summarize this chapter, Bolshevism, primarily in the figure of Lenin, remained true, by and large, to the views attributed earlier to Marx and Engels on the question of democracy—but only in theory. Bolshevism affirmed the class nature of both the struggle for democracy and ruling democratically. It maintained that democracy is truly realizable only in communism. However, Lenin saw this realization in the abolition of democracy—in this sense, he stood much closer to Engels than to Marx. What is new in Lenin is the conception of the vanguard party of professional revolutionary Marxists that stands above the pyramidal system of the Soviets and guides its conduct in order to assure its proletariat content. In the realm of practice, however, Bolshevism under Lenin failed to live up to the ideal of democracy put forth by the Paris Commune (and glorified by Marx and Engels). It also failed to actualize its Soviet version outlined by Lenin himself in his debate with Kautsky. The ideal of full-class dictatorship—and its companion ideal of the Soviet model of democracy—began to lose its attraction immediately following the victory of the October Revolution. The unfavorable objective conditions were coupled with low cultural levels to stand in the way of the realization of this ideal. Starting with the War Communism of the Civil War period, the party took over the control of the government from the Soviets and never gave it back. In explaining the failure of Soviet democracy in terms of the low cultural level of the Russian proletariat and peasants, one should note that although the two main figures of Bolshevism had grasped the depth of the problem, they were more concerned with the development of material and social culture than with political culture. In developing these cultures, they constantly looked to Western Europe and the United States and set out to emulate their ways. They wished to import these cultures from the West without importing the western political culture that accompanied them. (Lenin and Trotsky's vehement attacks on Western bourgeois democracy and political culture are well-known facts.[40]) By rejecting the Western model of democracy as a whole, the Bolsheviks were led to also reject what was good in its political culture.[41]

Thus, Bolshevism in power soon abandoned the true ideal of Soviet democracy—(that of establishing a system of the *direct participation* of ordinary people in governing themselves, or the ideal of the government *by* the people)—and in its place engineered a system of *indirect* rule by the people, or a government *for* the people by their most "advanced section," the party. This gave way to the rule of a bureaucracy that grew more powerful and aggressive, and eventually turned into the oppressive monster of Stalinism. Although the main focus of this study is on Lenin, it should be mentioned that Trotsky, the second leading Bolshevik, also had his own share of theoretical and practical failings on the question of democracy.[42]

In light of what was discussed above, conceptualizing Bolshevism as representing a somewhat coherent conception of democracy can only be justified in terms of Lenin's contention that the proletarian democracy can deliver to the people what bourgeois democracy promises, but fails to deliver, viz., *the political empowerment of the people.* Lenin's conception of the Soviet system of democracy presented in his counter-attack against Kautsky is truly a conception of the political empowerment of the common people. Had the Soviets worked according to their design, Bolshevism could have been characterized as a theory of government *by* the people, albeit not as a theory of fully direct democracy.[43] Moreover, Bolshevism can also be characterized as a theory of government *for* the people, for it did *intend* to function in this capacity, both in theory and in practice. The emphasis was placed on the *substance* (viz., the class-nature) of the policies the government initiated and laws it legislated, as well as being placed on the *righteousness* of *outcomes* it expected to achieve from these policies and laws—i.e., serving the class interests of the common people being the criterion of the righteousness. The negative side of Bolshevism as a conception of democracy was that it lost sight of procedures and legality that was revered in the West. Apropos of the question of being the government *of* the people, Bolshevism in practice falls short in the sense that its leadership did not sweep into power on the waves of massive national strikes or demonstrations, nor did it win its claims to legitimacy in popular elections.[44]

Notes

1. Macpherson (1965), p.13 and p.16.

2. Macpherson (1965), p.16

3. Macpherson does not emphasize the problems posed by the cultural backwardness of Russia.

4. Macpherson (1965), p.17.

5. The phrase "the free development of each is the condition for the free development of all" is taken from the *Communist Manifesto* (Marx and Engels 1977, Vol.1, p.127).

6. Macpherson's own view is that the countries in the so-called socialist bloc were democratic in a broad sense, i.e., democracy as equality (Macpherson 1965, p.22).

7. Lenin (1952), Vol.II, Part 1, p.303.

8. Ibid., p.304. Italicization of the world "all" is original; the rest is added.

9. Ibid., pp.242-44.

10. Ibid., p.248, original italics.

11. Ibid., p.249, original italics.

12. The debate started when Kautsky published a pamphlet titled *The Dictatorship of the Proletariat* in August 1918, where he advanced a series of criticisms against the new Soviet regime in Russia.

13. Lenin (1952), Vol.II, Part 2, p.48.

14. Ibid., p.49.

15. Ibid., pp.52-53, original italics.

16. Thanks to the new regime's seizure of the "best buildings" from the exploiters and turning them into meeting halls for the Soviets, ibid., p.55. One should add that contrary to what is commonly believed, the idea of "Soviets" is not originally a Leninist or Marxist idea. The idea gained popularity in Russia circa the Russian Constitutional Revolution of 1905. "The first Soviet [council] was formed in St. Petersburg in 1905" and Trotsky seemed to have played a major role in it, arguing that it was an "authentic democracy" (Patrick Goode in Bottomore 1983, p.96).

17. Lenin (1952), Vol.II, Part 2, p.55, original italics.

18. Ibid., p.56, original italics.

19. Ibid., p.55, original italics.

20. Ibid., p.56, original italics.

21. Ibid., pp.53-54, original italic. One should add that the views expressed by Lenin on democracy and the dictatorship of the proletariat in the debate are consistent with those expressed earlier in the *State and Revolution*.

22. Farber (1990), pp.210-11. Farber attributes this view to Ralph Miliband and Carmen Sirianni.

23. Ibid., original italics.

24. Ibid., p.211. A good example of Lenin's approach to this problem is his famous notes written on December 23, 1922, entitled *Letter to Congress* where he suggested adding new members "drawn from the working class" to the central committee of the communist party as a way of curbing the authoritarianism of the committee under Stalin's leadership (Williams 2000, p.194).

25. From 1905-June 1917, Lenin, according to Hal Draper, insisted that the dictatorship of the proletariat should be "an authority unrestricted by any law" (a view expressed by Lenin in 1905) and "power based not on law or elections, but directly on the armed

forces of a particular section of the proletariat" (a view expressed by Lenin in June 1917) (Draper 1987, p.90 and p.96, respectively). It should be added that in the *State and Revolution*, Lenin did not repeat this "no law" position and instead interpreted dictatorship as meaning "undivided power directly backed by the armed forces of the people" (ibid., p.96).

26. Draper (1987), p.104.

27. The main problems in the realm of practice had to do with the fact that the Soviets (which had mushroomed immediately before and after the revolution and had played an important role in seizing the political power) did not develop into the much-anticipated institutions of democracy either in the letter or the spirit of the Paris Commune. See Farber (1990) for a thorough discussion of the decline of the power of the Soviets under Lenin.

28. See Draper (1987) for a discussion of Lenin's theoretical inconsistencies on the question of the dictatorship of the proletariat, especially pp.93-105 and pp.133-37.

29. Trotsky (1973). Aside from the main evils of illiteracy, religious superstition, bribery, and alcoholism, Bolsheviks had to teach the Russian common people the "civilized" western ways and habits of life, such as "respectful speech," "politeness," "tidiness," reading, and the western habits of work such as "punctuality," "precision," "discipline," and "organization." In addition to the low levels of "material culture" and "social culture," some of the manifestations of which were just mentioned, the Russia of the time was virtually void of *political culture*. The educated middle-class and the intelligentsia in general, even the leaders of the revolution, had their own cultural shortcomings: arrogance and disdain for the people. "The condescension of the Bolshevik elite toward the Russian people" is a theme of Peter Kenez's *Birth of the Propaganda State* (1985), e.g., p.7.

30. Trotsky (1973), p.145.

31. In "On Co-operation," January 1923, in Lenin (1976), Vol.33, p.475.

32. In "Better Fewer, but Better," March 1923, in Lenin (1976), Vol.33, p.501.

33. Phrases "material culture and "social culture" are borrowed from Trotsky. See note 29 above.

34. In addition to being dominated by the old culture, the state bureaucracy, and increasingly the Party bureaucracy, were also being run by "the sons and daughters of petty officials of the old regime." As Bender explains, "[o]wing to the Bolsheviks' small numbers at the time of the October revolution, and to the fact that many Bolsheviks were killed fighting in the Red army during the Civil War (1918-1920), it became inevitable that the ranks of the expanding bureaucracy would become inundated with individuals who saw participation in the new order as the road to personal power and well-being. This was true of both proletarian and nonproletarain elements, as the bureaucracy and then the Party itself became swollen with the sons and daughters of petty officials of the old regime" (Bender 1975, p.50).

35. Draper mentions that Lenin in 1918 set out to write an "extensive" book on the dictatorship of the proletariat but never managed to do it (Draper 1987, p.136). One of the theses that he was going to develop in the books was "Dictatorship of the *revolutionary* elements of the class" (ibid., p.137, original italic).

36. In the "Report on the Party Programme" presented to the Eighth Congress of the R.C.P. (B), March 19, 1919, in Lenin (1976), Vol.29, p.183, italics added. Here Lenin was speaking within the context of fighting bureaucracy.

37. Farber (1990), p.197, e.g., in dealing with Mensheviks. In his study of Lenin's actions and writings in the early NEP years, Farber argues that "[while] Lenin inaugurated the NEP and thus relaxed state economic controls, in the political realm he moved from the very widespread but still somewhat tentative repression of the Civil War years to-

wards the complete and systematic repression of opposition parties and groups" (ibid., p.196).

38. Ibid., p.30.

39. Ibid.

40. In the *Problems of Everyday Life*, Trotsky repeated time and again that the strength of the Western European bourgeoisie lay in its wealth and culture (p.146). These enabled the Western bourgeoisie "to hold back the awakening [of] political self-determination of the proletariat" (ibid., p.146). It achieved this by instilling false consciousness ("conscious falsity," ibid., p.248) into the minds of the proletariat. The Western bourgeoisie, Trotsky believed, had perfected the "technique of lying" to the level of science, comparable to the "technology of electricity" (ibid., p.248). Bourgeois democracies, according to Trotsky, were based on "high-grade lying" and the "consciously organized deception of the people by means of a combination of methods of exceptional complexity" (ibid., p.248).

41. Here the allusion is to habits of voting, adhering to procedures, accepting defeats in elections, tolerating the views of opponents, etc. As one Soviet official commented some seventy years later,

> [w]e [in the Soviet Union] longed to find forms of democracy that did not continue the old forms but refuted them by demonstrating their bankruptcy. In the process we not infrequently threw out the baby—the common contents of democracy—along with the bourgeois bathwater. The worst of the traditions of old Russia filtered through into the new society via hundreds of different channels—psychological, political, and moral—and made themselves at home here.

(The comment was made by Fyodor Burlatsky, a Soviet official, *circa* 1988, in Brown 1990, p.47.)

42. For a discussion of Trotsky's "theoretical debacle" on these questions see Draper (1987), especially pp.139-41.

43. Not fully direct because Soviets had a pyramidal structure which started with the local Soviets at the base and moved up to the district congresses, and then to the regional congresses, and then finally to the All Russian Congress of Soviets which elected the Central Executive Committee of Soviets (CECS). This system was a fully direct democracy only at the base and delegatory democracy from the base up.

44. Recalling from Chapter 1, the phrase of government for/of/by the people is borrowed from Lincoln's "Gettysburg Address." Lenin dissolved the Constituent Assembly after Bolsheviks failed to win a majority there, arguing that Bolsheviks (with Left SRs) had the majority in the Soviets and that Soviets represented a higher quality of democracy than the one represented by parliamentary elections.

Concluding Remarks to Part I

The history of the idea of democracy in pre-liberal and non-liberal societies can be viewed as the history of the ideal of the political empowerment of the (common) people—or perhaps more fittingly as the history of the *hope* of making the (common) people sovereign in the state. In making this argument, Part I was principally guided by a main theme of Macpherson's *Real World of Democracy*. According to Macpherson, the idea of democracy in these societies had a class component; that is to say, the idea was coupled with a thick notion of equality. Broadly speaking, the idea of the political empowerment of the people or their sovereignty in these societies can be interpreted in two senses. In the first sense, the *political empowerment of the people* was understood as the people's *direct participation* in the legislative, executive, and judicial functions of the state, or as government *by* the people, *for* the people, and *of* the people. Ancient Athenian democracy was such a system, at least in principle. It understood equality primarily as the political-legal equality of all enfranchised citizens; this meant that every one was equal before the law and had an equal opportunity to take part in the state functions, regardless of his social or material standing.[1] Although this system appeared, or claimed, to be a non-class affair, in its actual practice, it was marked by class tensions that tilted the balance of power in the interests of the rich.[2] This in part contributed to the decline and the eventual demise of Athenian democracy.[3]

A variant of this sense of political empowerment is Rousseau's conception. Rousseau's understanding of sovereignty required the ordinary citizens' direct participation in decision-making. (It should be recalled that Rousseau also understood the people's participation in *ratifying* the laws made by their appointees as a form of direct participation.) One major difference between Rousseau's conception and ancient Athens' is that Rousseau was vehemently opposed to the people's direct participation in executive functions, as well as being opposed to their deliberation and forming factions or associations. In making up their own minds, i.e., in getting to discover the general will, the people would take their cue from a supreme legislator. The other major feature of Rousseau's conception that sets it apart from the ancient Athenian model is its insistence on the need for a thick, yet fluid, notion of material equality, in addition to the political-legal equalities the citizens of ancient Athens enjoyed.[4]

In its second sense, the idea of the political empowerment of the people, or securing their sovereignty, took the form of the claim, according to which, the hallmark of a democratic system was not that the people directly participated in ruling, but that their *true (class) interests* were sovereign in the state and served as its guiding principles. The fundamental criterion here was no longer the fact of the *direct* participation of the people, but that the ruling apparatus ruled with the *intention* of protecting and advancing their interests. This was the *distorted* version of the Soviet democracy that began to take shape under Lenin's reign. Unlike the first sense, this second sense declared democracy as fundamentally a class affair and prided itself in being the rule of the common people (or the poor). Had this system worked according to its original design—i.e., had the objective and subjective factors that contributed to its distortion been absent—it would have been a synthesis of the Rousseauean and Ancient Athenian conceptions in some of their ideal forms. That is to say, the common people (workers and peasants) under the political-cultural guidance and enlightenment provided by the communist party would have participated in the legislative, executive, and judicial functions of the business of governing and would have worked toward the abolition of class distinctions—albeit the forms of the people's participation in higher echelons of the governing-administering apparatus would have been indirect.

Moreover, in the pre-liberal societies of Athens, the Roman Republic, and the medieval cities, there existed no adequate legal-political institutions, nor any conceptions of universal rights or liberties, nor accommodating cultural attitudes, nor technological prerequisites needed for securing a moderately productive economy and prosperity that could justify providing all citizens with legal rights and other means that would empower them to take part in the franchise on a massive scale. Given these limitations, the idea of the *direct* participation of ordinary people could only be realized in small settings. Ancient Athens, the Roman Republic, and the medieval cities succeeded only because they limited the citizenship to a manageable size. Furthermore, the conception of liberty that existed in the pre-liberal societies of Athens, the Roman Republic, and the medieval cities of Europe were considerably different from the negative liberties guaranteed to the individual as a matter of social contract in the liberal society. The liberty of the pre-liberal society was a positive concept and was mainly intended to empower the individuals to fulfill their responsibilities as citizens. The scope of whatever individual liberties enjoyed by the citizens was subject to change and revocation. Liberal society's sanctifying and privileging of private life and negative liberties over and above the claims of the good of the community was an alien idea to the ancients.[5] Apart from his positive liberties as a citizen, the individual of the ancient city had no other liberties or rights known in the liberal society.[6]

Furthermore, given the low levels of the technologies of mass communication, mass transportation, and the construction of large assembly halls, not all citizens could attend the legislative or deliberative sessions.[7] More importantly, due to low economic productivity and underdeveloped technologies and modes of producing the material necessities of life, the common people (the poor) of the ancient and medieval cities—not to say anything about women, slaves, and peasants who were in bondage—could not afford the luxury of civic-leisure and time needed to educate themselves on political issues, and attend the democratic sessions, even if the accommodating cultural attitudes, appropriate social-political institutions, and physical spaces had existed for these purposes. In fact, the general economic prosperity that served as the economic contexts for the practice of democracy in these societies were achieved through exploiting the labor of women, slaves, and peasants—and also through collecting tributes from colonized people, especially in the case of Athens.[8] The same weaknesses are also true in the case of Rousseau's conception. It too was designed for a small and under-productive agrarian community (city-state or a canton) and thus was limited in its scope. Moreover, Rousseau's conception did not permit deliberations or associations, as it lacked faith in the intellectual capabilities of average citizens. Rousseau's solution to the problem of the "cognitive deficit" of the citizens was to establish the moral-intellectual authority of a supreme legislator who resorted to myths and religious deception as a way of intimating to the people their "general will."[9] Finally, Rousseau was more than willing to compromise the individual liberties and sacrifice them readily at the feet of the general will and the larger good of the community.

As to Bolshevism under Lenin, although the notion of individual liberties was not a completely alien idea in its domain, it was put on the back burner, and was trampled upon frequently. Some of the leading Bolsheviks, Lenin in particular, had lived in Western Europe and were quite familiar with the concept of negative individual liberties. However, knowing how the capitalist classes in the West used these liberties to justify their existence and safeguard their interests, and given Bolsheviks' well-justified suspicion that these liberties could endanger the survival of the insecure new Soviet regime, they forsook them whenever they felt threatened. The ultimate guiding principle of Bolshevism in dealing with the question of individual liberties under Lenin was the regime's imperative to prevent the infiltration and corruption of the Russian working-class movement by the bourgeoisie. Moreover, the Russia the Bolsheviks inherited from the czar was not only an economically under-developed country, but also lacked all of those cultural elements necessary for rapidly developing the kind of radical democracy and modern economy Lenin had envisioned on the eve of the October Revolution. Problems of everyday life "stood before" Bolsheviks "in all their concrete immensity."[10] The Civil War and the hostile attitude of the capitalist world toward the new regime made matters worse. This historical experimentation with democracy, too, lacked some of the essential elements needed to assure its success. Democracy was not going to flourish on Earth this time around either.

Its failings on the question of negative liberties aside, however, Bolshevism was on the same par with the liberal societies to the West, if not ahead of them, on the question of extending legal-political and socio-economic equalities to all citizens—of course, barring the members of the bourgeoisie.[11] Moreover, the elaborate pyramidal structure of the Soviets was, theoretically speaking, more than adequate to qualify it as a system of semi-direct participatory democracy or an effective system of control from below.[12] Lastly, a major theoretical flaw of

Lenin's conception of Soviet democracy was that it lacked a system of checks and balances and a "democratic procedure" for settling disagreements within the Soviet political system itself.

Notes

1. As Pericles put it in his funeral speech: "everyone is equal before the law: when it is a question of putting one person before another in positions of public responsibility, what counts is not membership of a particular class, but the actual ability which the man possesses" (Thucydides 1954, p.117). One should add that the ancient Greek political thought did not take for granted the natural equality or ability of all individuals, as was the case with Plato's. Moreover, one should be reminded that the Athenian democracy excluded the majority of the adult population of the city, viz., women, immigrants, and slaves, and thus the notion of equality did not apply to them.

2. Here, the allusion is to Pericles' funeral speech in which he claimed that Athenian democracy did not have a class bias: "Our constitution is called a democracy because power is in the hands not of a minority but of the whole people. . . . No one, so long as he has it in him to be of service to the state, is kept in political obscurity because of poverty" (Thucydides 1954, p.117).

3. Many other factors contributed to the fall of Athenian democracy. Among these, one should mention Athenians' cultural failings (oppression of women, slaves, and colonized people), growth of individualism and materialism among its citizens in the height of its colonial power, the rise of demagogues, the infiltration of oligarchs, and finally the conquest of Athens by Alexander the Great. The reader should consult Claster (1967) for a short overview of some of these factors.

4. Here, the allusion is to Rousseau's claim, considered earlier, to the effect that no one should be rich enough to buy another, and no one poor enough to be bought by someone else.

5. As Benjamin Constant put it: "All private actions were submitted to a severe surveillance. No importance was given to individual independence, neither in relation to opinions, nor to labor, nor, above all, to religion" (Constant 1988, p.311), in an address given by Constant to the *Athenee Royale* in 1819 in defense of the modern society against the ancient one). Also, "among the ancients the individual, almost always sovereign in public affairs, was a slave in all his private relations" (ibid.). However, John Gray argues that Constant's account of the absence of individual liberties among the ancients is exaggerated and that the "germs of liberal ideas" were present among the Greeks and were championed by the Sophists (Gray 1995, p.3). Similarly, Quentin Skinner believes that the preoccupation of the ancients (in the Roman Republic) with the ideal of "free government" can be taken to mean that the majority of ancients had a "fundamental desire to lead a life of personal liberty" (Skinner 1992, P.220). The latest scholarship seems to suggest that the "ancients" (in the Roman Republic) and the medieval European city-states (the "neo-Romans") saw freedom not in the positive sense, but rather in the sense of the absence of domination. Philip Pettit, for instance, argues that Constant was flatly wrong in claiming that the liberties of the ancients consisted of "democratic self-mastery" and that the idea of freedom that has dominated republicanism throughout its history is the "ideal of freedom as non-domination" (Pettit 1997, p.271). See also Philip Pettit (2002), and Quentin Skinner (1992). In light of these considerations, Constant's view on

the liberties of the ancients seems to have more relevance to ancient Athens than to the republican Rome.

6. It is worth noting that some (e.g., Gray 1995) have argued that "the germs of liberal ideas" had existed among ancient Athenians. Some sophists such as Protagaras have been assumed to hold liberal outlooks. A good example of such an outlook is to be found in Pericles famous Funeral Oration quoted earlier. Gray has argued that Rome had a tradition of "individualist private law", and political thinkers such as Cicero were essentially liberal thinkers (Gray 1995, pp.3-7).

7. In the case of Athenian democracy, one should be reminded that while the size of the enfranchised citizen body was estimated about 20,000-40,000, the amphitheater could seat only 6,000 people, albeit it rarely filled up (see Budge 1996, p.25).

8. Given its low technological levels, hence its under-productive mode of production, the economic prosperity that was needed for sustaining political stability in Athens, and thus its democratic franchise, had to come from exploiting the labor of the majority of the population (women and slaves), and from drawing benefits from the contributions made to the economy by non-Greek residents of the city. Tributes collected from colonies also contributed to the general prosperity of the city (e.g., paying for public works that employed poorer members of the city, and maintaining navy and army employing thousands of Athenians), thus contributing to the political stability needed for practicing democracy. (See Jones (1964), pp.3-20 for an opposite view. Jones is of the opinion that Athenian democracy could have existed without its slaves and colonies.) Although Athenian democracy paid citizens for attending the Assembly starting in 404-02 B.C., these pays were meager and could not have compensated fully for the time one would lose from work in order to become a full participant in the city's politics as well-to-do members, especially those of aristocratic birth, were. The poor generally stayed away, and as Jones speculates, they "probably preferred more profitable employment" (ibid., p.50). From the perspective of Historical Materialism, the underdeveloped modes and technologies of economic production in the ancient and medieval worlds were directly linked to their specific social-political institutions and cultural attitudes. Viewed from this angle, one should thus regard slavery and serfdom, and consequently, the absence of democracy on a massive scale, as inescapable facts of these worlds.

9. See the discussion in Chapter 3, especially notes 55 and 74.

10 Allusion to the statement quoted from Trotsky in note 30 in Chapter 5.

11. For example, Bolsheviks were enthusiastic supporters of full universal suffrage (including women's right to vote) and instituted the franchise right after the revolution. Women's suffrage in the United States had to wait until 1920. In the case of the United Kingdom, though women were given partial rights to vote in 1918 (conditioned on the age requirement of thirty or higher), the full suffrage had to wait until 1928.

12. This also seems to be Macpherson's evaluation of the Soviet system of democracy. By all indications, Macpherson did not believe that the Soviet system was inherently flawed (Macpherson (1977), p.109).

Part II

The Case of the Liberal State and Liberal Democracy

Chapter 6

Liberalism and the Rise of the Liberal State

In continuing to examine the history of the idea of democracy, Part II will be devoted to the study of the idea in liberal and liberal-democratic societies. The main guiding thread here will be the question of how modern Western societies in the nineteenth and twentieth centuries dealt with the ideals represented by democracy as they appropriated the idea and claimed to have made it the hallmark of their political systems. This examination will also be guided by inquiries into how these political systems addressed the questions of sovereignty, equality, and citizens' participation in political decision-making.

*   *   *   *   *

Starting in the nineteenth century, the idea of democracy in the West first had to succumb to, and then share the political center stage with a new and powerful political theory that appeared on the horizon in the seventeenth century. This new theory was none other than liberalism. What is often referred to as liberalism constitutes a wide spectrum of theories, views, and value-judgments on Man and society that began to take shape in the seventeenth century. Thomas Hobbes is often regarded as the forerunner of liberalism. Arguably, the rise of liberalism, and modern political philosophy in general, is much indebted to the Machiavellian revolution in the realm of political thought that re-introduced to early sixteenth-century Europe the core truth of "the political." The truth that politics was a this-worldly affair and had no divine connections—an idea well-known to the ancients—seemed to have been lost to the Christian world of the medieval period. Once the Machiavellian revolution consolidated itself, it was no longer possible to argue that the monarch had a divine right to rule; nor was it possible for him to rule with an absolute power. Moreover, no longer was it possible to argue that the business of ruling was about establishing a divine or moral order in the world. If absolute power could ever be justified, it could only be done so in the name of establishing order for the well-being of the members of society. This was the starting point of liberalism as it came into being with Thomas Hobbes—albeit Hobbes pushed the theory in a direction other than the Machiavellian brand of classical republicanism.[1] Hobbes' Leviathan was a this-worldly beast, created and supported by the members of society themselves, and in accordance with an original social contract agreed to by all. Given that the members of the Hobbesian commonwealth had given their *consent* to the rule of Leviathan freely, Leviathan could legitimately claim to be the government *of* the people.[2] Moreover, given that Leviathan delivered to the people their most desired good, viz., security, it could also lay a legitimate claim to being the government *for* the people, despite its oppressive and brutal rule.

However, the Hobbesian individuals agreed to surrender themselves to the political authority of an almighty and beastly power only because they themselves were beasts and therefore could only accept the sovereignty and authority of a beast more powerful than themselves. Given that the state of nature was a state of "war of all against all," a secure society under the rule of Leviathan was the most rational choice. Had the individuals agreeing to the contract had a benign nature, plausibly, they would have opted for a minimal government whose sole function would have been to adjudicate among them when their self-seeking endeavors came into conflict with the similar pursuits of others. This latter picture of the state of nature and the form of the state of society that it would naturally lead to was portrayed by John Locke. Save the extreme difference between taking Man in the state of nature as a violent beast, on the one hand, and as a benign creature, on the other hand—(and consequently save the diametrically opposed forms of governments they advocated)—both Hobbes and Locke held somewhat similar views on the attributes of Man. Both viewed the human individual in his natural state as fundamentally free, solitary, self-seeking, possessive, and (instrumentally) rational. Speaking solely within the context of the politics of their times, while Hobbes' liberalism was a reactionary one, in that he offered a rational argument for the absolute power of the status quo, Locke's, on the other hand, was a revolutionary world outlook. His liberalism proved to be a political map of the journey into the uncharted territory of the modern state and society that was inaugurated with the constitutional settlement and the Bill of Rights in England in the aftermath of the victory of the Glorious Revolution of 1688-89. These developments gave the British parliament supremacy over the crown as well as granting a legal status to the idea of the natural rights of Man. The victory of the Glorious Revolution marked the beginning of the era of liberal revolutions that lasted up to the mid-nineteenth century. Without a doubt, the greatest of these revolutions was the American Revolution of 1776-89, producing a constitution that translated some of the highest aspirations of liberalism into a system of fundamental laws.[3]

One should be quick to add that the rise of liberalism as a powerful political idea was accompanied by the rise of a new set of socio-economic realities, and thus by the concomitant rise of new socio-economic relations and the rise of a new economic class to the position of social and political prominence. A host of developments that started in post-Renaissance Europe—and lasted up to the mid-nineteenth century—brought about a new money economy which ended up requiring a new set of relations of production and distribution. The actual events and processes that contributed to this development ranged from the colonial plunder of the world by the Western European naval powers (which Marx later regarded as part and parcel of the "primitive accumulation of capital" in Europe) to the new technological innovations that improved the efficiency of production, transportation, and communication, as well as opening new areas of production and distribution.[4] The new rising economic class grew mainly out of the merchant and artisan sections of the middle classes of Western Europe. The rising mercantile capitalism, which was later joined by the fledgling industrial capitalism, demanded a new social-political and judicial-legal system that would recognize their new-found fortunes and properties, and would protect them against the whims and arbitrary decisions of monarchs and aristocrats. The new classes demanded new social and economic relations that were conducive to their economic growth. Thus, an epoch of social-political revolutions began that ended with the birth of the liberal state and society. To these, one should also add the historical fact that these revolutions were both assisted and encouraged by diverse spectra of revolutionary intelligentsia who found themselves in the midst of these upheavals. In this historical period, the class struggle of the rising bourgeoisie to win the right to enjoy the fruits of free markets was intertwined with the intellectual struggles to free reason from the yoke of faith, as well as being directly connected with the battle to win legal rights and liberties for the individual. Destroying the medieval state was the historical mission of liberalism and Locke proved to be its most significant missionary.

In its further development, which lasted up to the early-mid twentieth century, liberalism went through various phases of interpretations as it was shaped and reshaped at the hands of numerous and diverse figures ranging from Montesquieu, Adam Smith, and Edmund Burke to J. S. Mill and John Dewey. On the one hand, liberalism in its "classical" form was formulated as the political philosophy of the era of *laissez-faire* economics, as well as the ideological world outlook of capitalism. In this interpretation, liberalism postulated Man as an instrumentally rational, and self-seeking economic animal, and insisted on the sovereignty of the market, as well as insisting on Man's inviolable rights and liberties, while at the same time, championing the idea of limiting the power of government over the individual and the market. On the other hand, as a humanist world outlook, liberalism reflected the eighteenth-century Enlightenment belief in the dignity of the human individual and its commitment to reconstructing the world in accordance with the universal principles of reason, freedom, brotherhood, and equality. As a humanist world outlook, liberalism also reflected the humanism and philanthropic ideas flowing out of the reaction movement against the Enlightenment.[5] In the hands of Adam Smith and like-minded individuals, liberalism was shaped as a bourgeois political-legal movement that demanded, and achieved, freedom from governmental intervention in economic affairs. The *laissez-faire* liberalism of Adam Smith was reformulated by Jeremy Bentham on utilitarian principles. As will be seen later, this reformulation was later humanized, socialized, and "democratized" within a movement that started with J. S. Mill in England. The "new" liberalism of J. S. Mill, T. H. Green, and L. T. Hobhouse made major inroads into the intellectual consciousness of much of early twentieth century England.[6] In the United States, the new theory found its highest expression in John Dewey's works in the early to mid-twentieth century. In a span of almost half a century, liberalism went through a transition from a market-oriented political theory to a socially-oriented one. Its concerns with the question of maximal liberty in the marketplace, and its advocacy of limited government with which the theory preoccupied itself in its "classical" period, gave way to striving to reach a balance between its fixation on liberty, on the one hand, and the need to pay attention to questions of equality and social welfare, on the other hand. As will be seen later, this change of focus came about in part as a consequence of the loss of faith of many liberals in the ability of the market to regulate itself, and in part, in the aftermath of the success the state displayed when it intervened in the market in order to save it. This development was directly linked with the rise of the welfare state and thus the decline of the class-warfare state.

In closing this chapter, it is worth noting that despite the fact that liberalism went through numerous phases in its evolution as a political theory, and advocated somewhat of a wide range of social and institutional arrangements in the course of the last two and half centuries or so, as a political philosophy, it continues to rest on the same deep-seated beliefs and assumptions with which it began. The main body of these beliefs and assumptions, summarily stated, include the principles of individual freedoms, the moral and social primacy of the individual (individualism), the constitutional protection of civil liberties and rights, the moral and legal-political equality of citizens, the separation of private and public spheres, the belief in the sovereignty and independence of the individual in her private sphere of life, the belief in the moral and intellectual neutrality of the state, and finally the belief in separation of the state from society and economy, and thus, a preference for a small state.[7]

Notes

1. In making this claim, one should note the following. In a recent work, Philip Pettit has argued that the rise to prominence of the hallmark of liberalism, viz., the idea of negative liberty ("absence of coercion" or as "non-interference" as he calls it) did not begin with Hobbes, but with utilitarians Bentham and William Paley (Pettit 1997, p.45 and p.49). According to Pettit, "Hobbes' notion of freedom had little influence prior to the late eighteenth-century: that up to then, the republican notion of freedom as non-domination reigned more or less unchallenged in the English-speaking world" (ibid., p.44). Hobbes' notion of freedom was taken down from "the shelf of historical curiosities" in the late eighteenth century in order to be used as an ideological weapon against the desires of American colonies for independence (ibid.) This marked the ascendancy of liberalism which "displaced," as if in a "*coup d'etat*," the republican idea of liberty as non-domination which had enjoyed predominance in Western political thought up to that point (ibid., p.50).

2. According to Waldron, the "*fundamentally* liberal" idea is the thesis that "a social and political order is illegitimate unless it is rooted in the consent of all those who have to

live under it" (Waldron 1993, p.50, original italics). This is perhaps the main reason why Hobbes is regarded as being the forerunner of liberalism.

3. Three comments are in order here. First, the French Revolution of 1789, though it had a much greater impact and more historical significance than the American Revolution, cannot be regarded as an archetypal liberal revolution, for it was led by socialist and nationalist forces and ideologies as well as by liberal ones. Second, the Lockean elements in the American Constitution are clearly present in its recognition of the significance of the individual liberties and their protection by a government which is elected by, and accountable to, the public. The Hobbesian influence is present in its conception of representation that assumes politics originates in self-interest. As will be seen later in this part of the volume, it was Madison, the "main framer" of the American Constitution, who persuasively argued for this conception. Moreover, Montesquieu's influence is present, again through Madison, in the Constitution's strong principle of the separation of powers into the legislative, executive, and judiciary. Third, one should also add that some have argued that the American Constitution has strong republican ideals interwoven into its fabric. Michael Sandel, in particular, has argued that "[d]espite their revision of classical republican assumptions, the framers of the Constitution adhered to republican ideals in two important respects. . . . First . . . that the virtuous should govern, and that government should aim at a public good. . . . Second . . . that government has a stake in cultivating citizens of a certain kind" (Sandel 1996, pp.130-31).

4. The other parts and parcels of the primitive accumulation of capital were "forced labor, slavery, unequal exchange, colonial taxes" in colonies and "the traditional extortion of peasant surplus labor" (see Beaud 2001, p.79).

5. This is how John Dewey sketched the development of liberalism. To represent Dewey's viewpoint more accurately, one should add that Dewey believed that liberalism developed out of these two different streams as they "flowed into one," albeit "they never coalesced." (In the humanitarian camp, Dewey places Rousseau and the British Wesleyan movement among others.) Dewey lamented that it was the economic side of liberalism that prevailed in the movement (see Dewey 1975, pp.126-28).

6. According to John Gray, J. S. Mill's attitude toward trade unionism, nationalism, and "socialist experimentation," T. H. Green and Bernard Bosanquet's contentions against the then-prevailing conception of freedom, and L. T. Hobhouse's "synthesis of the philosophies of Mill and Green" began the movement that socialized and humanized liberalism (Gray 1995, pp.30-32).

7. Two notes are in order here. First, the main difference between the "classical" and "new" liberalism is that the sphere of citizenship in the latter is much wider and extends to all segments of society. Moreover, as a consequence of this extension, the principle of political-legal equality applies to all. Thus, all acquire the opportunity to enjoy the fruits of liberty on an equal basis, hence "liberal democracy"—and this is what is meant above by liberalism becoming "democratized." Second, *here, liberalism is taken mainly as a broad doctrine or a social-political world outlook that, taken from a narrow "historical" perspective, revolves around the advocacy of individual liberties as the primary social value—and not as a spectrum of socio-political views and positions that are often contrasted with "conservatism" and "libertarianism" in the contemporary politics of North American liberal-democracy.* (As will become apparent in the ensuing pages of this volume, this interpretation of liberalism will be contrasted with *democratic ideals*, which together with liberalism represent the two opposing poles of the liberal-democratic formula.) When Ronald Dworkin takes "a certain conception of equality" as being the "nerve of liberalism," he takes liberalism in its contemporary sense in North America

(Dworkin 1985, p.183). (Dworkin goes on to identify the "constitutive political morality" of liberalism, i.e., "the liberal conception of equality," as the one that is "neutral on . . . the question of the good life" (ibid., p.191-92).) Dworkin summarizes these "liberal positions" as being "for greater economic equality, for internationalism, for freedom of speech and against censorship, for greater equality between the races and against segregation, for a sharp separation of church and state, for greater procedural protection for accused criminals, for decriminalization of 'morals' offenses, particularly drug offense and consensual sexual offenses involving only adults, and for an aggressive use of central government power to achieve all these goals" (ibid., p.181). Jeremy Waldron, on the other hand, takes a broader understanding of liberalism when he states that "it is clear enough that a conviction about the importance of individual freedom lies close to the heart of most liberal political positions" (Waldron 1993, p.37). Waldron argues that, both in politics and in the sphere of private life, liberals "raise banners" of individual freedoms (freedom of intellectual pursuits, speech, association, religion, life-style, sexual practice, etc.) and civil liberties (ibid., p.38). Against Dworkin's view that these freedoms are derived from the liberals' neutral conception of equality ("equality of concern and respect"), Waldron goes on to argue that, "it seems to me that equality of respect, at least, cannot be understood in this context except by reference to a conviction about the importance of liberty (for everyone)" (ibid., pp.38-39). For Waldron, what is *"fundamentally* liberal" is the thesis that "a social and political order is illegitimate unless it is rooted in the consent of all those who have to live under it" (ibid., p.50, original italics). Now, returning to the original contention that here liberalism is taken broadly as a doctrine that mainly revolves around the value of individual freedom, it can be argued that the contemporary conservatives and libertarians can also be regarded as liberals, i.e., as *fundamentally* liberals. What separates "the liberals" (i.e., the liberals in the lexicon of contemporary North-American politics) from the conservatives is that the former have adopted a much broader and stronger notion of equality than the conservatives—that is to say, they have become socialized, humanized, and "democratized." (Another difference between the two is that the conservatives' conception of equality, as Dworkin argues, does not grant complete neutrality to the state on the question of the good life (Dworkin 1985, p.191ff). Conservatives are preoccupied more with the enforcement of "public morals," traditional values, and concerns for "desert" than with concerns for equality.) Finally, the present-day libertarians are also *fundamentally* the old classical liberals who have revolted against the socialization and democratization of liberalism that began in the latter part of the nineteenth century and produced "the liberals." (Here, one can also use the contrast drawn by Will Kymlicka between the "right-wing libertarianism" (e.g., Robert Nozick's) and "left-wing liberal egalitarianism" (e.g., Rawls and Dworkin) as constituting the two opposing ends of "procedural liberalism" (Kymlicka 1998, p.132).)

Chapter 7

The Liberal State and the Idea of Representative Government

General philosophical assumptions of liberalism aside, insofar as the questions of governing and political authority and legitimacy are concerned, liberalism put forth forcefully the idea that only a government elected by the people—in other words, only a representative government of the people—could be regarded as a legitimate government. Without a doubt, this idea was a revolutionary one and proved to be the main building block of both liberal political theory and the liberal state.[1] Given its historical significance, and the powerful presence it has in the contemporary world, the modern idea of representative government deserves special attention and careful examination. This will be the focus of the rest of Part II.

To begin with, given that the very idea of modern representative government rests on *the value-claim that only an elected government could be legitimate*, the claim itself turns into a foundational principle for the idea. For this reason, this claim will be treated here as "the main principle" of representative government. Now, from the beginning in the liberal state, "the main principle of representative government" was accompanied by a set of four other (accompanying) convictions and assumptions that were interconnected with the idea—be it in theory or in practice. These convictions and assumptions were of great importance and value to the idea of representative government. And for this reason, they too will be regarded as "principles" of representative government. Taken together, these five principles will be treated in this work as laying the groundwork for both the theory and practice of representative government.

Among these four accompanying principles, one was the conviction that the individuals who would be at the helm of the representative government, in other words, the individuals who would represent the people, would be *superior* to the people themselves, be it in wealth or social standing, or in intelligence or education. Another presupposition here was that the functions of legislating and political decision-making in the representative government would be performed collectively and in accordance with some established procedures. (As will be seen later, this principle followed from the main principle.) The third among these accompanying principles was a specific understanding of the nature of representation that rendered the people politically passive. Finally, the last principle in this group mandated that the representatives be subjected to frequent and competitive elections. Given that these five principles (including the main principle) were, and continue to be, of fundamental importance to the modern idea of representative government, and also given that they represent the main features of this idea, in the rest of this volume, the idea of representative government will be conceptualized as resting upon these principles. In what immediately follows, these principles will first be defined in concise terms, and then will be discussed at some length.

Starting with the "*main principle of representative government*," this principle can be stated as holding that *only a freely elected representative government can be regarded as a legitimate government*. This idea is distinctly an "invention" that belongs to the *modern* world.[2] And Hobbes is the individual who is often credited as being its inventor.[3] One should keep in mind that despite its oppressive nature, Leviathan was a representative government.[4] Although Hobbes' own preference for the form of representative government was the government led by a single individual with absolute power, the form of representative government that came to dominate liberal theory early on was a government that was composed of assemblies, and had more than one branch. Now, given that there would be more than one representative, and thus given that representatives would have to work cooperatively with one another, the idea of representative government would necessarily become intertwined with the idea of making decisions collectively, and in accordance with some set of rules and procedures. This idea will be referred to here as the "*principle of collective-procedural decision-making*."

Moving to the next principle, the idea that the representatives ought to be "superior" to those whom they represent, will be referred to here, following Bernard Manin, as the "*principle of distinction of representatives*."[5] The principle of distinction requires that the representatives of the people be distinct individuals; that is to say, the representatives ought to be superior to their constituents, be it in wealth, in status, or in talent or character. Unlike the "principle of collective-procedural decision-making" that follows as a corollary of the "main principle of representative government," the principle of distinction of representatives has no theoretical connections to the main principle. Rather, given that the liberal society was a class-divided society, the principle of distinction came about as a matter of practical necessity for securing the domination of the privileged classes.[6] The fourth principle here is the idea that regards the business of representing the people as a function that would be performed independent of the wills of the latter, and without their intervention. This principle will be referred to here as the "*principle of independent-virtual representation*."[7] Finally, what will be referred to here as the "*principle of frequent and competitive elections*" is the idea that the representatives of the people would be elected via the mechanism of frequent and competitive elections, an arrangement in which the candidates would compete against one another in order to earn the right to represent the people. What follows is a detailed discussion of the last four principles.

*   *   *   *   *

First the "principle of distinction of representatives." There exists extensive historical data to substantiate the claim that, from the beginning in the liberal state, it was taken as a matter of a self-evident truth that only the distinguished members of society should be eligible to serve as the representatives of the people.[8] In other words, it was understood that the representatives had to have "higher ranks" than the people. Either they had to belong to a class of institutionalized or hereditary aristocracy (in the case of England), or they had to be distinguished from others by their wealth, social status, or talents (in the case of the United States). In short, they had to belong to an elite group or to an aristocratic class, be it hereditary or a naturally formed one—a "natural aristocracy."[9] In order to make sure that this principle was secured, various schemes and arrangements, and property qualifications, as well as legislative measures were employed. To substantiate that the principle of the distinction of representatives indeed was at work in the liberal society, one can turn to Bernard Manin's overview of the development of representative government in his *Principles of the Representative Government*.[10] Manin's work examines the rise of representative governments in England, France, and the United States. In the case of England, according to Manin, in the seventeenth and eighteenth century, "membership in the House of Commons was reserved to a small social circle."[11] As Manin explains, one factor contributing to this was a cultural one: "voters tended to take their cue from the most prominent local figures and considered it a matter of course that these prominent figures alone could be elected to the House of Commons."[12] A second factor was "the exorbitant cost of electoral campaigning, which increased steadily following the civil war [the Glorious Revolution of 1688] and throughout the eighteenth-century."[13] Finally, in 1710, "a formal property qualification was then established for MPs," which was different and "higher" than the one required for the electorate.[14]

In the case of France, and again according to Manin, the idea that the representatives must be distinct from the people was first realized by legislating laws that specified property qualifications. The Constitutional Assembly in 1789 required that "only those who could meet the two conditions of owning land and paying taxes . . . equivalent of 500 days' wage . . . could be elected to the National Assembly."[15] However, in the face of opposition, the Assembly in 1791 repealed this hefty property requirement and resorted to a system of indirect elections that produced the same results. As Manin notes, the resulting assemblies continued to be "dominated by the wealthy classes."[16] This situation in France did not change until 1848 when the manhood suffrage and direct national elections were introduced.[17]

However, in the case of the United States, the principle of distinction was realized "in a more flexible and adaptive manner" in comparison to how it was achieved in England and France.[18] Given that the American Constitution could not produce a property qualification for representatives, albeit against the intention and desire of its framers, the principle of distinction in the United States had to be secured by indirect means.[19] This failure was mainly due to the fact that there existed considerable differences in the income and property levels in the country at the time.[20] In order to go around this obstacle and ensure that the representatives in the House would turn out to be a distinct cast of individuals, the Philadelphia Convention and the ensuing ratification debates managed to accomplish two things. First, the U.S. Constitution ended up stipulating *large* electoral districts for the House of Representatives. Second, the Federalists managed to establish a certain conception of representation that in combination with large electoral districts would greatly increase, if not guarantee, the likelihood that only distinct types of individuals would get elected. In order to see how exactly these developments came about, one needs to revisit the point of contention between the Federalists and the Anti-Federalists.

The Anti-Federalists argued for the view that the representatives must "resemble" or "mirror" the people whom they are elected to represent, or that they must be "like" their constituents in order to truly represent their views or sentiments. For them, this "resemblance" and "likeness" requirement was the sufficient condition for genuine representation—frequent elections being regarded as the necessary condition.[21] Against this conception, James Madison, the main framer of the U.S. Constitution, argued to the effect that "the representatives may well be different from the people, indeed they *ought to* be different."[22] The representatives would be, in Madison's own words, "men who possess most wisdom to discern, and most virtue to pursue, the common good of the society."[23] Moreover, they would constitute "a chosen body of citizens, whose wisdom may best discern the true interest of their community and whose patriotism and love of justice will be least likely to sacrifice it to temporary or partial considerations."[24] While rejecting the Anti-Federalists' notion of the "resemblance" of the representatives to the constituents as the sufficient condition for "good representation," the Federalists, most notably Madison, argued that frequent elections would assure good representation, and thus would be "both necessary and sufficient condition" for it.[25] Alternately stated, for the Federalists, the notion of the resemblance of the representatives to the people not only was irrelevant to the question of good representation, but was also wrongheaded; for the attributes of wisdom, virtue, and the love of justice were assumed to be absent among the ordinary people. If the representatives were to resemble the people, it then would follow that they would also be deficient in these attributes.[26] In short, as Manin puts it:

> The Federalists . . . all agreed that representatives should not be like their constituents. Whether the difference was expressed in terms of wisdom, virtue, talent [Madison], or sheer wealth and property [Hamilton], they all expected and wished the elected to stand higher than those who elected them.[27]

One should add that the Anti-Federalists did not all come from a common socioeconomic background and were not of the same political disposition or of the same mind on the question of representation.[28] What united Anti-Federalists was their opposition to what they regarded as an utterly "aristocratic" conception of representation that the Federalists were advancing. Anti-Federalists, according to Joshua Miller, "were conservatives who defended democracy."[29] That is to say, the Anti-Federalists stood for the "localist," direct-participatory and community-based forms of politics that was "a way of life" in the America of the seventeenth and eighteenth-century.[30] Having stated this, however, one needs to add that their preference for participatory forms of politics should not be seen as an indication that the Anti-Federalists were true or full egalitarians, or that they were in favor of the participation of all segments of society in the political life. Nor should this be taken to mean that they were opposed to the then-prevailing social or economic inequalities. On the question of representation, Anti-Federalists, as Manin puts it, "wished only that the main components of society be represented," and placed "a special emphasis on the middling ranks."[31] In other words, they too favored a limited franchise that only included propertied white males (as opposed to favoring the institutionalization of the white male universal suffrage).[32] The Anti-Federalists also agreed with the Federalists that the representatives had to be "superior" to the enfranchised population they were expected to represent.[33] For them, the controversy was only a *quantitative* matter. They did not favor too great of a difference or distinction between the elected and electors. In the ensuing ratification debates, the controversy centered on the suggested ratio of representatives to electors. The Philadelphia Convention had set the ratio of one representative for every thirty thousand inhabitants, thus producing a House of Representatives of the size of sixty-five members.[34] The Anti-Federalists argued for a higher ratio, and thus for a larger House of Representatives. This meant that they were arguing for smaller electoral districts. In smaller districts, they contended, middle-class citizens such as "professional men, merchants, farmers, mechanics, etc." could also stand a chance of being elected.[35] It should be emphasized that the Anti-Federalists were not arguing that the *ordinary* middle-class citizens should have the chance to be elected, but only that "*their best informed men,*" i.e., the individuals of highest rank among them, should have the chance to represent the views and sentiments of their fellow middle-class citizens.[36] Despite their differences on the ratio of representatives to the represented, and consequently over the size of electoral districts, both Federalists and Anti-Federalists agreed that all electoral districts must be of the same population size. At the end, it was the Federalists' conception of representation with its large electoral districts that prevailed as the Federalists ended up winning "the battle over the Constitution" against the Anti-Federalists.[37]

It should be noted that Madison, despite using the language of "virtue," "wisdom," "love of justice," "patriotism," and the "common good" in defining the attributes of representatives—(e.g., in "The Federalist No. 10" and "The Federalist No. 57," quoted above)—agreed that large electoral districts were "manifestly favorable to the election of persons of general respectability, and of probable attachment to the rights of property" and would "work in favor of property and wealth."[38] Madison's own admission aside, the representation in the United States did become a prerogative of those who were attached to wealth and property—

though not necessarily in the earlier decades of the new republic—and in practice worked in the same manner as it did in Europe.[39] The experience of the United States showed that the principle of the distinction of the representatives could also be realized without imposing property qualifications or indirect elections, and that the desired result could be achieved by relying on the "mere operation of the elective method."[40] As Manin is quick to point out, it took Europeans almost a century to give up their reliance on property qualifications or legislative measures, and mimic the method of Americans.[41]

Finally, in closing the discussion of the "principle of the distinction of representatives," one should add that the idea of paying representatives salaries, as was mandated by the U. S. Constitution of 1787, succeeded, against the original intention of the framers, in curbing the aristocratic character of representation in the House and enabled the less well-off middle-class members of the society to be elected to the House.[42] This, one should be quick to add, did not necessarily break the attachment of the representation to wealth and property. What it did instead was to allow interests associated with the middle-class types of wealth, property, and professions (small tradesmen, freeholders, farmers, artisans, etc.) to find representation in the House alongside the "landed interests," "commercial interests," "moneyed interests," and other major interests that were generally associated with what Anti-Federalists referred to as the "well-born few" and "natural aristocracy"—whose interests they believed the Federalists sought to advance.[43] In the case of England and France, given that property qualifications for representatives were much higher than those for the electorate, the idea of compensating the representatives for their services did not serve as a vehicle for enabling the less well-off—(who were not even eligible to vote)—to join the ranks of representatives.[44]

    *   *   *   *   *

Now, the examination of the "principle of distinction of representatives" leads directly to the discussion of the "principle of independent-virtual representation." As was evident in the American debate, these two principles were interconnected. In this debate, the questions of the eligibility of representatives and the size of electoral districts were closely tied to the questions of what, or whom, exactly, the representatives were expected to represent, and how the function of representation was expected to be performed. Alternately stated, the questions were as follows. Should the representatives represent "interests," or should they represent "persons"? Assuming that they were expected to represent "interests," should they think globally with the interests of the entire nation in mind, or should they think locally within the confines of their own electoral districts? Moreover, assuming that they were expected to represent "persons," should they merely transmit the wills and wishes of the persons they represent directly to the House, or should they go by their own conceptions of what these wishes were and what would need to be done about them? In the American scene, although these questions hung in the balance, they did not come to the forefront of the ratification debates. One can speculate that the main reason for this was that there seemed to be a general understanding about the function of representation in the Philadelphia Convention and among those who participated in the ratification debates.

Nowhere is this general understanding more clearly present than in the contributions of Madison to the debates. Roughly stated, in this understanding of the function of representation, the main emphasis was not so much on representing the *interests* of the whole nation or the common good, as the representatives understood them to be, but mainly on representing *persons*, i.e., the persons *who had interests*. In other words, the framers of the U.S. Constitution understood the function of representation to consist in representing the interests of the constituents as the constituents themselves understood them, and not as the representatives perceived what these interests were or what they ought to be. However, this does not mean that in this conception of representation, the representatives would be required to take instructions or mandates from their constituents on what their interests were. Nor would the constituents have the power to recall their representatives in case the representatives failed to represent the constituents as they saw fit. Nor was it the case that the common good or interests of the whole nation had no role to play in the decisions to be made in the House of Representatives. Quite the contrary, the opposite was true in both cases.

In the general understanding of representation that prevailed in the American debate, the representation was not conceptualized as a transmission mechanism for carrying directly the wills or wishes of the people to the House. Moreover, as was mentioned above, the representatives were not expected to be instructed by their constituents; nor were they legally obliged to take instructions from them; nor could they be recalled by the people for failing to legislate in accordance with their will, or for failing to carry through what they had promised in their campaigns.[45] (As a matter of fact, the right to instruct representatives was debated in the first Congress as part of the Bill of Rights, and was rejected.) The voters, it was understood, had the freedom to instruct their representatives, but the representatives were not legally bound to abide by them.[46] Insofar as the question of keeping the representatives attuned to the constituents' own interpretations of what their interests were, the constituents could only rely on the threat that they would not reelect their representatives if they failed to represent them as they wished. Madison emphasized this point repeatedly.

Thus, the prevailing understanding did not conceptualize representation as an *actual* or literal one. On the contrary, it took representation as *virtual*, and the representatives as *independent* individuals who would represent not the fleeting desires of their constituents, but what was constant in them. In other words, what the representatives would bring to the table in the House, would not be the raw desires, opinions, or purely self-regarding interests of their constituents, as the constituents express them, but rather the refined and processed versions of these interests. Now, when it came to legislating, in addition to promoting the interests of their own constituents, the representatives also had to be concerned with how the interests of their constituents could be *balanced* by the interests of other constituents in other parts of the country—which could be different from theirs or even be in conflict with them. It seems that the overriding concern in the Madisonian conception of representation, the one that prevailed in the American debate, was that the representatives would be able to establish a balance among the plurality of competing, and even conflicting, interests that existed throughout the original thirteen states that were coming together to form a new union. For a federal republic to take root and survive, it was essential that an equilibrium among the diverse interests of these states, and among the constituents as well, could be reached. Given the importance of reaching equilibrium, the American representatives had to have the attributes of virtue, wisdom, and impartiality argued for by Madison; and this is why only an elite and distinct group of individuals, a "natural aristocracy" indeed, could qualify to serve as representatives.[47]

These superior qualities of the representatives were essential to the task of compromise-making, as well as being handy in preventing the conflicts among opposing interests from spinning out of control. The framers of the U.S. Constitution, especially Madison and Hamilton, were keenly aware of the differing class interests and class conflicts that existed among the enfranchised classes in the United States and they often spoke of the conflicts of interest among the "landed interest," "manufacturing interest," "mercantile interests," "moneyed interest," "creditors," and "debtors." [48] And Madison, in particular, "almost always" spoke of interests "in the plural."[49] For this reason, Madison's conception of representation is directly related to the question of factions, and ultimately to that of the class conflict that arises from "an unequal distribution of property."[50] Given this wide field of competing and conflicting interests, Madison's conception of representation can be regarded, "in a wider sense," as a mechanism for, as Hanna F. Pitkin has argued, "bringing dangerous social conflicts into a single central forum, where it can be controlled by blanching and stalemating." [51] A similar account of what representation was expected to achieve in the Madisonian conception is offered by Melissa Williams. According to Williams, Madisonian representation assumes that, given the large multiplicity of competing interests, no particular interest would be able to prevail as the majority interest in the House, thus compelling the competing interests to seek compromises. As she puts it, "the competing interests in Congress would cancel one another out . . . [and] . . . only those programs that are in the common interests of the whole will command a majority of the representatives, and in this way the public good will be preserved."[52]

*   *   *   *   *

It should be noted that the debate over the question of what or whom the representatives should represent had started in England about a century before it was debated in the United States.[53] Unlike the United States, the sense that had prevailed in England was that representation was not about representing *persons*, i.e., the persons with interests, but about representing the *interests*, and in that, the interests of *the whole nation*.[54] Almost a decade earlier, in a famous speech given in 1774, Edmund Burke had argued persuasively against the conception of actual representation or representation as delegation. The elected officials of the people, according to Burke, should not be allowed to legislate laws based on the will of their constituents, nor do they have any mandate to do so. As the representatives of their constituents, the elected officials are not their deputies or "agents" and "advocates" against the advocates or agents of a contending set of interests or adversarial groups; nor are they the transmitters of their constituents' decisions. So far from all of these, the representatives, according to Burke, are *independent* political actors who assemble "only with *one* interest" in mind, that of the *whole* nation, and who, as the *trustees* of the people, legislate for "the general good, resulting from the general reason of the whole."[55] The speech was given after Burke was declared a winner, to be one of the representatives of the city of Bristol. The relevant part of the actual passage is as follows:

> Parliament is not a *congress* of ambassadors from different and hostile interests, which interests each must maintain, as an agent and advocate, against other agents and advocates; but parliament is a *deliberative* assembly of *one* nation, with *one* interest, that of the whole—where not local purposes, not local prejudices, ought to guide, but the general good, resulting from the general reason of the whole. You choose a member, indeed; but when you have chosen him, he is not [a] member of Bristol, but he is a member of *parliament*. If the local constituent should have an interest or should form an hasty opinion evidently opposite to the real good of the rest of the community, the member for that place ought to be as far as any other from any endeavor to give it effect.[56]

By all indications, the conception of representation put forth by Burke is also elitist. The representatives in his view form, in the words of Hanna Pitkin, "an aristocracy of virtue and wisdom."[57] They represent reason and right, and not the wishes or wills of their constituents. Burke takes for granted that the electorate are narrow-minded, self-seeking, and incapable of seeing the good of the whole community; whereas the representatives, on the other hand, are superior individuals who rise above their constituents' self-seeking impulses; and guided by wisdom and virtue, discern the interests of the whole nation, upon which they "deliberate" and legislate. Moreover, this conception of representation has a message of class compromise and the cessation of class-warfare. Not only should the representatives not act as transmitters of the will of their constituents, and instead cast their votes in the legislative assemblies according to their own consciences, but also, they should not let the particular interests of their constituents guide their thinking on the issues. As the trustees of the entire nation, the representatives should legislate with the interests of the whole nation in mind. In England, no doubt, this second aspect of Burke's conception of representation greatly facilitated the expansion of the franchise of democracy to the lower classes which, as will be seen in the following chapter, ensued almost a century later. If the members of the lower classes could be persuaded to accept this notion of representation, then they could be allowed entry into the franchise, for this would prevent the parliament from turning into a class-war zone, and would leave the privileges of the propertied classes unchallenged.

As can be seen, there is a considerable difference between Burke's and Madison's views on who or what is to be represented in the House of Commons or in the House of Representatives. Burke's conception opts for representing the interests of the whole nation—supposedly even those who have not yet been franchised. Madison's conception, on the other hand, opts for representing persons who have interests, i.e., only those who have been enfranchised, and are powerful enough to have their voices heard. For Burke, the interests of the whole nation are "objective," "impersonal," "unattached," "broad," "fixed," and "few in number"— and apparently this is the main reason why only the representatives are capable of understanding and representing them.[58] As Pitkin puts it, "Burke almost never speaks of an individual's interest, or of the interests of a group."[59] For Madison, on the contrary, the interests were personal, attached, "subjective," "pluralistic," and constantly "shifting."[60] However, despite these differences, both conceptions were elitist and aristocratic, and both attributed to representatives qualities and skills that could enable them to do what ordinary people were assumed to be incapable of. "Both believed that," as Williams puts it, "institutions should be structured so as to encourage the election of a *talented elite* inclined toward the public interest."[61] By all indications, while the Madisonian conception of representation is realistic in many respects, Burke's is highly idealistic. Burke's notion appeared even more idealistic by the advent of liberal democracy that, as will be seen later in this part, multiplied the number of "legitimate" interests. Moreover, Madisonian representatives were less removed from ordinary people, and their real-life situations, than Burkean representatives. And perhaps this is the main reason why the Madisonian notion of representation survived the test of time, and Burke's died away.[62]

*   *   *   *   *

The discussion of the differences between Burke's and Madison's view of the function of representation brings to a close the discussion of the "principle of independent-virtual representation"; and leads directly to the discussion of the "principle of collective-procedural decision-making." As has become evident by now, both Burke and Madison agreed that the decisions in the House of Commons and House of Representatives would be made collectively by the representatives and in accordance with some established procedures. For Burke, the guiding principle for policy and decision-making, and thus the underlying principle for establishing procedures and rules that would produce these policies and decisions, was the conviction that the decisions reached in the House should result from the *deliberation* of the representatives who would subject the issues to the scrutiny of wisdom, reason, and moral considerations, with the interests of the whole nation in mind. Viewed from the perspective of Madison, such an approach to the question of decision-making would appear as purely utopian. As A. H. Birch has argued, "[t]he starting point of Madison's analysis was his belief that a conflict of interests in society is inevitable and will necessarily lead . . . to the development of factional disputes on political questions."[63] Given this inevitability of the conflicts of interests and factions, the guiding principle of decision and policy making, and thus that of devising procedures and rules for collective decision or policy making, must be the commitment to balancing the conflicting interests. In Madison's understanding, collective decision-making is the process of engaging in an optimal process of negotiation, compromise, and bargaining among the elected representatives or officials who represent competing, self-regarding, and yet pliant preferences and interests of the persons or groups with interests. The goal of collective decision-making was to combine, or to relate to one another, individual choices, preferences, and interests in ways that an equilibrium among the competing interests would emerge, or an optimal satisfaction would be achieved.

Finally, a brief discussion of the "principle of frequent and competitive elections." The notion of frequency of elections was an essential component of the modern idea of representative government since the beginning in the liberal state. In the American debate, recalling from the earlier discussion, frequent elections were regarded by the Federalists both as a necessary and sufficient condition for good representation. It was assumed that this principle would assure that the representatives would not drift too far apart from their constituents and that the latter would manage to exercise some control over their representatives, albeit indirectly. As to the competition aspect of the "principle of frequent and competitive elections," it should be mentioned that given the fact that non-propertied classes were excluded from the franchise (hence a small size of eligible voters), and in light of the fact that the voters had a good knowledge of where their interests lay, competition among the candidates for office was a relatively low-key affair. Seldom was there real electoral competition in districts when it came to the question of interests. In the electoral districts where landed interests predominated, candidates represented the interests of the landowners. In the same vein, the candidates running in predominately manufacturing and commercial districts represented manufacturing and commercial interests respectively. Given that the non-propertied and working classes were not franchised, the candidates for the House did not take up campaign slogans or issues that were of interest to them, which could have clashed with the big interests in their districts. As will be seen later in this part and Part III, once the non-propertied classes were admitted into the franchise, the competition aspect of the "principle of frequent and competitive elections" took on a new meaning and a greater importance.

*　　　*　　　*　　　*　　　*

To bring this chapter to a close, liberalism as a theory of government and political authority began as a theory of representative government. From the beginning, the liberal state had a representative form of government. Moreover, the idea of representative government, both in theory and practice, was an elitist and aristocratic construct. It was designed to admit only a select and special type of individual into the government. It should also be noted that, in practice, the idea of representative government turned out to also be a class-based conception. It attended to the interests of those who had material stakes in society and excluded those who had minimal or no stake in the system, the majority (the people). The elitist character and aristocratic nature of the idea of representative government in the liberal state is clearly evident in the "principle of distinction of representatives" and the "principle of independent-virtual representation." The representatives were not assumed to be just "a chosen body of citizens" in the sense that they have been elected, but also that they constituted, as Manin has argued, a "chosen Few" who were superior to the rest of the citizens, for they were wise, virtuous, and could rise above the level of the "howling masses" or the "swinish multitude" on matters of self or sectional interests and could think and act "impartially."[64] Moreover, the representatives were assumed to be independent and virtual representatives who either set their agenda in the House independent of the wills of their constituents (Burke), or took to the House the "filtered" versions of the wills expressed by their constituents (Madison).[65]

Now, returning to Abraham Lincoln's phraseology of "government *by* the people," "government *of* the people," "government *for* the people" in his famous Gettysburg speech, one can argue that the Madisonian government cannot be characterized as being of the first two types. The people (the majority, i.e., the non-propertied individuals, women, slaves, and Native Americans) were completely left out of the picture. Madisonian government cannot be regarded as the government *by* the people, for the people were not in it; nor can it be considered as the government *of* the people, for the people had not elected it. As to being a government *for* the people, the Madisonian government could qualify only minimally. Although it was designed to represent only those who had interests, the Madisonian government was also to serve the "true interests of their country," albeit indirectly.[66] As is the case with Madison's, Burkean government also fails to be the government *by* the people or *of* the people for the same set of reasons. However, given its claim that it intends to work for the whole nation, even those who are disenfranchised, the Burkean government has a more legitimate claim than the Madisonian one for being the government *for* the people.

Notes

1. According to John Gray, nineteenth-century Europe, especially nineteenth-century England, was the "golden age of liberal theory and practice" (Gray 1985, p.27). He estimates that this period stretched between the years 1815 and 1914 (ibid., p.32).

2. Hanna F. Pitkin has argued that the idea of representation is "essentially a modern one" (Pitkin 1967, p.2). According to Pitkin, although the ancient Greeks and Romans elected officials, they did not think of the elected ones as their representatives (ibid., pp.2-3). As to the medieval period, the idea of representation gradually rose in the thirteenth and fourteenth century in feudal England. According to Anthony H. Birch, "by the end of the fourteenth-century representative institutions were part of the machinery of government in over a dozen states" (Birch 1971, p.29). Having mentioned this, it should be stressed that representation in medieval Europe was not recognized as a "matter of right," as Pitkin puts it (Pitkin 1967, p.3). Rather, it was a matter of privilege that was often dependent upon land ownership. In the case of England, as Birch contends, "[w]hen the king summoned his magnates and bishops to a royal council he did so because they were men of power, and when he extended the summons to include knights and burgesses

he did so because they could speak on behalf of other property-owners" (Birch 1971, p.29). "Nevertheless," as Birch concludes, "it is from these bodies that the representative institutions of the modern world are descended" (ibid.).

3. According to Robert Scigliano, "[t]he doctrine of representation was the invention of the English philosopher Thomas Hobbes, who used the term in his *Leviathan* (1651). For Hobbes, representative government was a "government authorized by the people" (Scigliano in Lipset 1995, Vol.3, p.1056).

4. In the words of Birch, "Leviathan was in fact the people's representative, having been established in that capacity at the time of the original contract" (Birch 1971, p.32). In Hobbes' own words:

> It is manifest, that men who are in absolute liberty, may, if they please, give Authority to One man to represent them every one; as well as give such Authority to any Assembly of men whatsoever; and consequently may subject themselves, if they think good, to a Monarch, as absolutely, as to any other Representative (Hobbes 1996[1651], Chapter XIX, p.130).

Hobbes thought of representation as "Soveraignty by Institution," and compared it with the case when "the Soveraign Power is acquired by Force" (ibid., Chapter XX, p.138). What separates Hobbes' notion of representation from the modern conception of the idea is that his is completely devoid of any notion of *accountability*. The representative or the "Soveraigne Instituted" cannot be accused of injustice, cannot be punished for his actions, nor can he be put to death by his subjects, for his subjects are the "Author of all the Soveraigne doth; and consequently he that complaineth of injury from his Soveraigne, complaineth of that whereof he himselfe is Author; and therefore ought not to accuse any man but himselfe" (ibid., Chapter XVIII, p.124).

5. Manin (1997), p.94.

6. As will be seen later, the principle of distinction grew out of the debates around setting eligibility criteria for the representatives in major countries wherein the liberal state first took hold, viz., England, the United States, and France.

7. As will become clear later, the main motivation for establishing this principle was to protect the institution of private property.

8. One should be quick to point out that by "the people" here, it is meant only the enfranchised segments of the population.

9. The phrase "natural aristocracy" is borrowed from Manin who uses it to separate the representatives from the rest of the population as a distinct class of individuals. Manin himself borrows the term from Anti-Federalists in the American constitutional debates, in particular, from Melancton Smith and Brutus (Manin 1997, p.113).

10. Manin (1997). Despite what the title of Manin's work might suggest, the idea that representative government rests upon five principles, as was conceptualized above, is original to this book. One should further add that the principle of the distinction of the representatives here almost entirely relies on Manin's own principle of distinction. Moreover, as will become evident in the rest of this chapter, the author is greatly indebted to Manin's work in formulating and presenting the concept of representation.

11. Manin (1997), p.95.

12. Ibid., p.96.

13. Ibid., p.97.

14. Ibid., p.97.

15. Ibid., p.99.

16. Ibid., p.101.

17. Birch (1971), p.63.

18. Manin (1997), p.129.

19. As it turned out, the Philadelphia Convention of 1787 did not succeed in determining a fixed property qualification that could be used uniformly throughout the original thirteen states. Thus, the Constitution of 1787 stipulated that the eligibility for voting in federal elections were to be the same as those required by the states for their lower houses. As the historian Howard Zinn has stated, at the time, "almost all the states" required property-holding qualifications for voting (Zinn 1980, p.95). It should be noted that these qualifications applied only to white males, thus excluding women, blacks, and Native Americans.

20. According to Manin (1997), the authors of the U.S. Constitution *did* intend to include a property qualification for representatives but could not achieve it: "the absence of property qualifications in the 1787 constitution was not due to the reason of principle, but of expediency. . . . [T]he exceptionally egalitarian character of representation in the United States owes more to geography than to philosophy" (ibid., p.107). Moreover, it should be added that in addition to the difference of income within the various parts of the geographical territory (mainly South vs. North), there was another problem that stood in the way of requiring a fixed and uniform set of property qualifications. This problem was, as the committee that was commissioned to specifying a property qualification reported, "the difference between the present and future circumstances of the whole, rendered it improper to have either *uniform* or *fixed* qualifications" (Manin 1997, p.106, italics added, Manin quotes historical records).

21. The notion that representatives must "resemble" or "mirror" their constituents is often characterized as a "descriptive" or "microcosmic" theory or conception of representation by some authors, e.g., by Williams (1998), p.28, and p.326, and Pitkin (1967), p.79, p.81, and p.87.

22. Manin (1997), p.117, italics added.

23. Madison in "The Federalist No. 57," quoted by Manin, ibid., p.116.

24. Ibid., p.117.

25. Ibid., p.118.

26. This is based on the assumption that the representatives were expected to resemble the *general* or *common* attributes of the people, which were assumed to be devoid of wisdom, virtue, etc. The main problem with the resemblance theory of representation is that it fails to specify, as Melissa Williams puts it, "*which* attributes of citizens especially deserve to be represented" (Williams 1998, p.29).

27. Manin (1997), p.121.

28. Historian Gordon S. Wood considers some of the most prominent Anti-Federalists (e.g., George Mason and Richard Henry Lee) as "aristocrats," while regarding others as "plebeians" (Wood 1992a, 258-259). Wood also considers the aristocratic Anti-Federalists as "republicans" (in the same sense that he considers Federalists of Madison's stature as republicans), and the plebeian Anti-Federalists as "localist" or "populist" (ibid., also see Wood 1992b, p.99 and p.103). There are those who consider some of the leading Anti-Federalists (e.g., Sam Adams and Patrick Henry) among the "real democrats" of the American Revolution (e.g., Barber 1998, p.82). Beitzinger considers George Mason, Patrick Henry, and Richard H. Lee, George Clinton, Robert Yates, Eldridge Gerry, and Luther Martin of Maryland among the "prominent men"—(or using John Adams' terms, the "gentlemen")—in the Anti-Federalist camp, and the rest primarily "simplemen"

(Beitzinger 1972, p.214). Miller regards the Anti-Federalists mainly as "conservative" democrats (Miller (1991), p.82). Most authors consider the Anti-Federalists as being more republican than democratic, for they stressed the importance of teaching civic virtues and practicing self-government on local levels, and believed that they too supported federalism, but argued that the federal government had to be small and simple (e.g., Wilson Carey McWilliams in Lipset 1995, Vol.1, pp.69-70).

29. Miller (1991), p.82.

30. Ibid., p.132. Miller traces the practice of the Anti-Federalist style of democracy in the America of the time to the Puritan theology and traditions that were brought to America in the earlier years of the colony. In Miller's account, what laid at the heart of the Federalist agenda was an attempt to establish a "set of institutions that embodied the more undemocratic aspects of liberalism" as a "substitute for democracy" that had already established itself as a tradition in the America of the time (ibid., p.133). In somewhat similar ways, Duncan sees "Anti-Federalism as an extension of a much older, much larger historical/political context that not only predates the ratification debates over the Constitution but also, in metamorphosed form, actually postdates those debates as well" (Duncan 1995, p.128). Duncan believes that the Anti-Federalists' conception of democracy "was a more classically inclined democratic vision of small, face-to-face democratic city-states, or small republics" (ibid., p.129).

31. Manin (1997), p.112.

32. As Barber notes, the "real democrats [of the American Revolution] (Sam Adams, Patrick Henry, Tom Paine, Jefferson himself) were not present at the Philadelphia Creation, and [this is perhaps why] radical democratic models calling for unicameral legislature and universal white male suffrage of the kind represented by the Pennsylvania Constitution were given short shrift" (Barber 1998, p.82). One should also add that the universal suffrage was not fully realized in the United States until 1964 (the year of the Constitutional Amendment xxiv) when the poll tax qualification was dropped, and blacks in the South were finally allowed to vote. The Constitution of 1787 had stipulated that the eligibility for voting in federal elections was to be the same as that required by the states for their lower houses. It is a well-established fact that almost all states at the time had imposed tax or property qualifications for eligibility. These qualifications applied only to white males, thus excluding women, blacks, and Native Americans (Zinn 1980, p.95). By 1825, the white male suffrage was in place in nearly all the states (Wood 1992b, p.101). The expansion of the federal franchise began in 1870 when the franchise was expanded to all males (Constitutional Amendment xv in 1870). The next major step was the enfranchisement of women in 1920 (Constitutional Amendment xix). The franchise became fully realized by the passage of the Constitutional Amendment xxiv in 1964.

33. Manin (1997), p.133.

34. Ibid., p.107.

35. Ibid., p.131. The phrase quoted above has been taken from the passage quoted by Manin from *The Federal Farmer*, Letter II. The "Federal Farmer," according to Wood, put forth this view as a "fair representation"—Wood goes on to characterize this view of representation as an "extreme form of actual representation" (Wood 1992b, p.101).

36. Manin (1997), p.131. The phrase quoted above has been taken from the passage quoted by Manin from *The Federal Farmer*, Letter II. The italics are added by Manin.

37. The phrase inside the quotation marks is borrowed from Wood (1992a), p.259. One should add that some (e.g., Wood) have argued that despite the fact that Federalists won the battle over the Constitution, and put in place what Anti-Federalists considered to be an "aristocratically designed [political system that was intended] to 'raise the fortunes

and respectability of the well-born few, and oppress the plebeians'," it was the Anti-Federalist conceptions of representation and government that prevailed in America for almost a century following the ratification of the Constitution (Wood 1992a, p.255 and p.259). Similarly, Miller argues that the aspirations of the Federalists were not fully realized "until after the Civil War" (e.g., Miller 1991, p.140).

38. This point is made by Manin (1997), pp.123-24. The phrases inside quotation marks are Madison's and have been taken from his "Note to the speech on the right of suffrage." The note in question was "an elaboration on the speech he delivered at the [Philadelphia] Convention on August 7, 1787." According to Manin, the note was written probably in 1821 (ibid., p.123).

39. One should note that the (House) representatives, despite their attachments to property and wealth, did not turn out to be mainly of the aristocratic types ("virtuous" and "wise") coming from the social strata of the landed gentry or "moneyed men" that Madison or Hamilton had hoped for (and the type that Anti-Federalists had feared would "raise the fortunes and respectability of the well-born few, and oppress the plebeians"). On the contrary, they turned out to be mainly "politically ambitious men" of the middle-class stock who pursued "interest-group" and "localist" politics and brought respectability to the idea that representatives "were elected to promote the particular interests and private causes of their constituents"—of course this stood diametrically opposed to the Madisonian "classical republican ideal that legislators were supposed to be disinterested promoters of a public good" and individuals of high moral "virtue," "wisdom," and "impartiality" (see Wood 1992a, p.255 and Wood 1992b, p.103). Paying representatives salaries helped the House to attract the individuals of middle-class origin who promoted mainly middle-class concerns. The fact that individuals of this disposition and social standing found their way into the House, and not the Federalist "well-born" types, could be taken as an indication that it was not the Federalist "moneyed men" and "landed gentry" that eventually won the political power and earned the most wealth in the century that followed the American Revolution, but the middle classes or "Antifederalist or Republican entrepreneurs." This interpretation, as well as the contention that the American government and politics went the Anti-Federalist way after 1787, is advanced by Wood. (See Wood 1997, pp.146-47, Wood 1992a, pp.258-59, and Wood 1992b, pp.91-105.)

40. Manin (1997), p.130.

41. Ibid.

42. Paying salaries to representatives was stipulated in Article I, Section 6 of the U.S. Constitution. According to Wood, starting in the 1780s, the social-cultural climate of the country began to change in ways that challenged the then-prevailing tradition of "classical republican thinking" which held that the individuals who served the public should not be paid. The argument now was that the ancient attitude deprived the ordinary people from serving the public and made the public service an exclusive prerogative of the "leisure" class (i.e., the landed gentry and individuals of similar social status who were wealthy enough not to work for a living, and thus could go to the public service without expecting compensation). Thus the old republican attitude began to appear as an utterly unegalitarian idea. (See Wood 1992b, pp.104-105, and Wood 1997, pp.146-48.) This social-cultural climate appears to be a main reason for having Section 6 appear on Article I.

43. See notes 37 and 39 above. See also Manin (1997), p.112-113, and Wood (1992a), p.255. Anti-Federalists, according to Manin, were more interested in representing the "middling ranks" of society. Paying salaries to representatives worked to the benefit of the Anti-Federalist view in that it allowed professional men, merchants, traders, and oth-

ers to find their way to the House. As to the class origins of the differences between the Federalists and Anti-Federalists, Beitzinger notes that "areas of subsistence farming tended to be Anti-Federalist, whereas commercial areas and areas of farmers dependent upon commercial markets tended to be Federalist." He goes on to add that "in the ranks of the Anti-federalists were men of means and men of commerce, as well as men of 'middling' and low ranks," and that "there were many 'gentlemen' and 'simplemen' on both sides, with a proportionally greater number of the latter among the Anti-Federalists" (Beitzinger 1972, p.214).

44. In the case of England, the idea of paying salaries to representatives was fiercely resisted until 1911. However, the debate against the idea continued as late as the late nineteenth century. The prevailing sentiment in England was that doing so would make representation, as J. S. Mill put it, "an object of desire to adventurers of a low class" (Mill 1958 [1861], p.170).

45. This went against a deep-seated belief that had established itself in the American colonies very early on, e.g., in the Massachusetts settlement. Wood reports that "in the eyes of most patriots the instructing of representatives had become 'an undoubted right'. And several of the states explicitly provided for this right in their new constitutions" (Wood 1969, p.190). Moreover, "many Americans believed their representatives to be . . . mere agents or tools of the people who could give binding directions 'whenever they please to give them' " (ibid., p.371).

46. Manin (1997), p.164.

47. One should add that Michael Sandel attributes to Madison's conception of representation, especially to its elitism, a republican character. Sandel argues that admitting private or sectional interests was a scheme invented to let the virtuous and the wise rule with the interests of the public "beyond the sum of private interests" (Sandel 1996, pp.130-31). In order to distance Madison from "modern-day interest group pluralists," Michael Sandel argues that "[f]or Madison, the reason for admitting interests into the system was not to govern by them but to disempower them, to play them to a draw, so that disinterested statesmen might govern unhindered by them" (ibid., p.131). Wood also considers Madison and other elite Federalists as "republican" (e.g., Wood 1992b, pp.98-9 and p.103), and puts the blame for the rise of interest group pluralism and politics in the United States on the shoulders of the Anti-Federalist's "localist" version of politics and the "explicit form of representation" (Wood 1992a, p.259). See also note 37 above.

48. These interests are mentioned by Madison in "The Federalist No. 10" (Hamilton 1961, p.131).

49. Pitkin (1967), p.192.

50. Madison in "The Federalist No. 10" (Hamilton 1961, p.131). It goes without saying that the conception of representation that prevailed in the U.S. constitutional debates presupposed a class-divided society. And it is no secret that the framers of the U.S. Constitution were keen on protecting the privileges and rights of property-holders against the possibility of incursion by the representatives. The framers were class-conscious and wealthy men who had numerous wealthy associates who were also class-conscious property-holders, and feared that the democratic fervor of the period could jeopardize the security of private property (e.g., the attacks on property in the state legislatures were growing in intensity at the time). This view was put forth strongly for the first time by Charles Beard in his *Economic Interpretation of the Constitution* (1935[1913], e.g., p.152-56). Separating the Declaration of Independence from the Constitution of 1787, Beard advanced the view that the latter was, to use the phraseology of Appleby *et al.*, "a reactionary document calculated to blunt the genuinely democratic forces unleashed by

the Revolution" (Appleby 1994, p.138). Beard's interpretation amounted to the argument that "the framers were actually antidemocratic . . . [and that they] . . . were less concerned with constructing a democratic government than with protecting economic interests" (Woolley 2002, p.65). (A concise overview of the critique of Beard's interpretation is found in Slonim (2001).) Following in the footsteps of Beard, some have argued that the main purpose of the system of checks and balances and the protection of minorities that the framers of the Constitution insisted on was to make the subversion of the institution of private property as difficult as possible, and thus keep the then-existing system of sharp class inequalities intact (e.g., Baran and Sweezy 1968, p.157). How well Madison understood the importance of class conflict and class-analysis in politics is evident in the following statement of his that reads as if it has been taken out of the opening pages of *The Communist Manifesto* which would not appear until sixty years later: "Those who hold and those who are without property have ever formed distinct interests in society" ("The Federalist No. 10," Hamilton 1961, p.131). The framers of the U.S. Constitution recognized the highly probable possibility that the property-less majority could overwhelm the political system that they were about to design by their sheer numbers, and endanger the security of the institution of private property. One should be reminded that the Philadelphia Convention was called in part as a response to the dangerous situation that Shay's Rebellion in the summer of 1786 had created, and in part to address the need for designing a national system for the expansion of commerce and industry, as well as for the protection of credit and property in general. How important were the considerations of the security of private property in the minds of the framers is illustrated by Jennifer Nedelsky's analysis of the thoughts and designs James Madison—the most important among the framers and "the Father of the Constitution"—had in mind for the Constitution at the time. "In 1787," according to Nedelsky, "the importance of the rights of persons and the rights to participate in making the laws were *assumptions in the back of Madison's mind*. The protection of property [on the other hand] was *the objective he held steadily before him* as he worked on the Constitution" (Nedelsky 1990, p.66, italics added).

51. Pitkin (1967), p.195. It should be noted that the presentation of the general understanding of representation that prevailed in the U.S. Constitution (i.e., Madison's view), as was done above in the last two pages or so, was based on a reading of Chapter 9 of Pitkin (1967).

52. Williams (1998), p.39. In other words, this "frustration of particular interests" (ibid., p.42) would channel "the forces of interest to produce the common good" (ibid., p.40) that would be acceptable to all. In Williams' view, relying solely on the "politics of factional competition" (ibid., p.42) is inadequate for producing "any substantive common good" (ibid.). According to Williams, what Madison's conception lacks is "a politics of deliberation over the common good" (ibid.).

53. Birch discusses in passing the debate between Tories and Whigs in the seventeenth century. While Tories were of the view that "the function of M.P.s was to represent local interests and to seek redress for particular grievances," Whigs believed that "Parliament was a deliberative body, representing the whole nation" (Birch 1971, p.38). Here Birch is quoting Algernon Sidney, a Whig, who expressed the Whig attitude on the question in the late seventeenth century (ibid.).

54. A direct result of this understanding of representation in England had been to prohibit the practices of giving representatives instructions or mandates or recalling them—this prohibition was adopted in the American and French constitutions (Manin 1997, pp.163-64). Moreover, that the voters could influence their representatives only by the

threat of not reelecting them was also emphasized in England. As Bentham argued later, "voters should only be allowed to influence their representatives by their right not to reelect them" (Manin 1997, p.164). J. S. Mill also supported the prohibition of the practice of giving representatives instructions (see Cunningham 2002, p.90).

55. Burke's position on the function of representation was not a novel one. As Birch puts it, Burke "was . . . following in a long tradition" in England (Birch 1971, p.39).

56. Burke (1901), Volume II, p.96, original italics.

57. This is how Pitkin (1967) characterizes Burke's thought on the question, p.172.

58. These are Pitkin's characterizations of interests in Burke (Pitkin 1967, p.168 and p.174).

59. Ibid., p.174.

60. Ibid., p.193.

61. Williams (1998), p.40, italics added.

62. Burke's notion of representation, as was presented above, captures only the dominant sense of representation in his thought. According to Pitkin, Burke was not always consistent and there were some other minor strands of thought present in his conception of representation. For instance, although he thought that the constituents' wills or particular interests were irrelevant to legislatures, he believed that their "feelings" and "sentiments" should be "reflected" or "reproduced" in representation (Pitkin 1967, p.183). Moreover, he believed that the particular interests of the Irish Catholics and American colonies should be represented (ibid., p.173).

63. Birch (1971), p.79.

64. Manin (1997), p.117. The phrase "a chosen body of citizens" was used by Madison in "The Federalist No. 10" where he gave his definition of republic. The following is the relevant passage in "The Federalist No. 10": A republic is "the delegation of the government . . . to a small number of citizens elected by the rest. . . . The effect of [which] is, on the one hand, to refine and enlarge the public views by passing them through the medium of a chosen body of citizens, whose wisdom may best discern the true interests of their country and whose patriotism and love of justice will be least likely to sacrifice it to temporary and partial considerations" (Hamilton 1961, pp.133-34). Moreover, the phrase "howling masses" was used by Hamilton to characterize the people (Barber 1998, p.81). Furthermore, the phrase "swinish multitude" was used by Burke. As was noted earlier in note 47 above, Michael Sandel regards Madison's claim that the chosen few would be wise and virtuous, and would govern in accordance with the interests of their country as an indication that he, along with other framers, "adhered to republican ideals" (Sandel 1996, pp.130-31).

65. As is evident in "The Federalist No. 10," the filter was the "medium of the chosen body of citizens": "to refine and enlarge the public views by passing them through the medium of a chosen body of citizens" (Hamilton 1961, p.134).

66. Ibid.

Chapter 8

The Rise of the Liberal-Democratic State

Although the liberal revolutions in Europe were won mainly by the participation of the peasants, the proletariat, and urban plebeians who flocked to the ranks of revolutionary movements, the liberal societies and states that were eventually erected on their corpses proved to be, by and large, the exclusive properties of the new propertied classes. The new liberal society operated on the basis of the economic relations of the competitive market and the social-political superstructures that were laid down largely by the market. Hand in hand with this infrastructure went the political-legal system of the liberal state that not only sanctified and supported it, but also modeled itself after the competitive and individualistic characteristics of its free-markets. The guiding principles of both the liberal state and liberal society were "freedom" and "choice," or just "freedom of choice" and "free choice." Although the availability of freedom of choice was a blessing—especially when it was contrasted with the all-pervasive bondage of the old feudal regime—the problem with the new society and state was that the new economic class, the bourgeoisie, had more actual and potential choices than the rest of society, and that it was far freer than the non-propertied classes who constituted the majority of the population. The new propertied classes, in cooperation with the old ones, had managed to establish a complete dominance over the "economic society," and consequently, over the "political society," the state.[1]

Despite the formal equality of all, the non-propertied classes of the new society were greatly disadvantaged. This state of affairs was a source of numerous problems for the liberal society from the moment of its inception. The liberal state not only served the free-market, but also modeled itself after it. C. B. Macpherson's statement of this problem in his analysis of the liberal state in *The Real World of Democracy* is most illuminating. As he puts it,

> the government [of the liberal society] was put in a sort of market situation. The government was treated as a supplier of certain political goods—not just the political good of law and order in general, but the

specific political goods demanded by those who had the upper hand in
running that particular kind of society. What was needed was the kind
of laws and regulations, and tax structure, that would make the market
society work.[2]

But how did this state work? And what was wrong with the way it worked insofar as the question of democracy is concerned? Macpherson's portrayal of the
political life in the liberal state answers these questions so eloquently that the
whole passage is worth quoting at length.

> The way was of course to put governmental power into the hands of
> men who were made subject to periodic elections at which there was a
> choice of candidates and parties. The electorate did *not* need to be a
> *democratic* one, and as a general rule was not, all that was needed was
> an electorate consisting of the men of substance, so that the government
> would be responsive to their choices.
>
> To make this political choice an effective one, there had to be certain
> other liberties. There had to be freedom of association—that is, free-
> dom to form political parties. . . . And there had to be freedom of
> speech and publication, for without these the freedom of association is
> of no use. These freedoms could not very well be limited to men of the
> directing classes [the propertied classes]. They had to be demanded in
> principle for everybody. The risk that the others would use them to get
> a political voice was a risk that had to be taken.
>
> So came what I am calling the liberal state. Its essence was the system
> of alternate or multiple parties whereby governments could be held re-
> sponsible to different sections of the class or classes that had a political
> voice. There was *nothing necessarily democratic* about the responsible
> party system. . . . The job of the liberal state was to maintain and pro-
> mote the liberal society, which was not essentially a democratic or an
> equal society. The job of the competitive party system was to uphold
> the competitive market society, by keeping the government responsive
> to the shifting majority interests of those who were running the market
> society.[3]

This situation of blatant inequalities in political participation and representation could not go on for long. The liberal state was grappling with the question of
legitimacy and was facing the threat of instability under the weight of the demands
by the disadvantaged classes who, to use Macpherson's words, "had no vote" and
thus "no political purchasing power" and "their interests were, by the logic of the
system, not consulted."[4]

Under the pressure of demands, and as a way of stalling the revolutions that
were sweeping through Europe during the mid-to-late nineteenth century, finally,
the liberal state gave in. And once it gave in, it was obliged to follow through with
the logic of equality that it itself had unleashed: it gave in to the demands of the
lower classes to have a voice in it. *Thus, the liberal state adopted democracy.*

So finally the democratic franchise was introduced into the liberal state. It did not come easily or quickly. In most of the present liberal-democratic countries it required many decades of agitation and organization. . . .

So democracy came as a late addition to the competitive market society and the liberal state. . . . It is not simply that democracy came later. It is also that democracy in these societies, was demanded, on competitive liberal grounds. Democracy was demanded, and admitted, on the ground that it was unfair not to have it in a competitive society. It was something the competitive society logically needed. . . . From a threat to the liberal state it [democracy] had become a fulfillment of the liberal state.[5]

This is why the liberal state adopted democracy, and *this is how the liberal-democratic state came into being,* first in Western Europe (England in particular), and later on in the United States in its own unique way and form.[6] By all indications, the adoption of universal suffrage (or the franchise of democracy) by the liberal state was not entirely a voluntary decision.

Once the lower classes were enfranchised, and gained representative power and voice in the state, they demanded, and received, various services such as education, welfare, and health benefits. Welfarism was a birthmark of the liberal-democratic state. As Macpherson notes, the "liberal-democratic state has typically become a welfare state, and a regulatory state. The rise of the welfare and regulatory state is generally held to be the result of the extension of the franchise."[7] Another development that accompanied the expansion of the franchise was that the trade-unions of the newly enfranchised classes won recognition and respectability from the state. It is worth noting here that from the Bolshevik standpoint, these concessions were carefully crafted by the bourgeois classes of Western Europe in order to install the working-class revolutions and save the system of exploitation from destruction. The working-classes of the Western European countries, as Lenin and Trotsky repeated time and again, were deceived by their bourgeoisie, and under the leadership of the individuals of Kautsky caliber, were led to betray their ascribed and historical interests.

In the beginning, the propertied classes resented the idea of having the lower classes represented in the state. In England, a considerable amount of political and intellectual effort and utilitarian calculations went into trying to minimize the effects of the participation of the lower classes in voting, as well as into trying to manage the devastation that the participation of the "swinish multitude" could cause.[8] The members of the lower classes were considered not only ignorant, uneducated, uncultured, and lacking a stake in the system, but also dishonest and untrustworthy. In England, in particular, some advocated schemes of voting that were intended to minimize the effects of granting voting power to the lower classes. The most popular of these schemes that were eventually implemented was the property qualification for the eligibility to vote.[9] In order to make sure that the lower classes would not be able to storm the liberal state and take it over in one blow, the extension of voting rights was done gradually. The process of admitting the lower classes into the franchise took place over a period of fifty years or so.[10] And as time went on, and as it was proven in practice that the lower classes (despite the fact that they constituted the greatest majority of the population) would not be able to take over the state—(thanks to the shrewd policies of the bourgeois politicians, the gradual pace of enfranchisement, and the old strategy of divide and conquer)—the resentment of the propertied classes changed to enthusiasm. Once the danger of instability and the fear of takeover by the "swinish multitude" had passed away, the liberal state congratulated itself in finding a great legitimization tool in democracy.

For the liberal state, democracy turned out to be a lifesaver and a blessing at the same time. In the way it was adopted, democracy ended up saving and strengthening the market society, rather than destroying it. It seems as if it worked according to a well-designed scheme. In the words of Macpherson, "liberal-democracy . . . was brought into being to serve the needs of the competitive market society."[11] No longer was democracy perceived as a "bad thing," or as something to be feared. The transformation of the concept from a "bad thing" to a "good thing" then began. And by the time President Woodrow Wilson declared that "the world must be made safe for democracy"—(as a way of justifying the United States' intervention in World War I)—it was apparent that democracy had won, as Macpherson puts it, a "full acceptance into the ranks of respectability."[12]

*   *   *   *   *

Thus, "democracy was liberalized," and was put in the service of the market society as Macpherson has shown.[13] However, this development came at a great cost to the idea of democracy. This liberalization could only be achieved by perverting the original meaning of democracy. The conceptions of democracy that predominated in the pre-liberal and non-liberal societies—("rule by the people," the empowerment of the people in the state, the sovereignty of the people in the state, and the implications of these conceptions, such as the direct popular participation or a substantive conception of equality, as were discussed in Part I)—were abandoned in favor of settling for the notions of universal suffrage, the competitive and choice-based conception of politics, and formal legal-political equalities. Thus, the meaning of democracy was redefined and reinvented; or rather, a new meaning was invented for it. The invention of a new meaning for democracy was necessary in order for it to be accepted as a category of the principles of "freedom," "choice," and "right" that had already been woven into the very fabric of the liberal market society and its state. This is how the perversion of the idea of democracy in the modern era began and this is how the public consciousness of the liberal-democratic society began to associate, and equate, democracy with freedom. Compared to the liberal society, the liberal-democratic society—(or the "free world," as it was branded later)—is democratic *only in that* it grants to all of its adult members the equal legal right to vote in elections, as well as the formal opportunity to be eligible to run for government offices. This is the only substantive difference between liberalism and liberal democracy on the question of governing and political authority-legitimacy.

The other principles that served as the foundations of liberal democracy on the question of governing and political authority-legitimacy were the same as the five principles of representative government considered earlier in discussing the liberal state. (As was argued, these principles were: the "main principle of representative government," the "principle of collective-procedural decision-making," the "principle of distinction of representatives," the "principle of independent-virtual representatives," and the "principle of frequent and competitive elections.") Liberal democracy inherited these principles from the liberal state and society. What is new in the liberal-democratic state is that the domain of eligible voters or electors has grown from a select group in the liberal state (in the beginning only the propertied and adult-native-male citizens) to the entire population. Therefore, the main democratic feature of the liberal-democratic society is but the criterion of the eligibility of all citizens to vote, as well as their eligibility to run for government offices, regardless of their social and economic standings—*universal suffrage*. (It is worth noting that in many liberal-democratic societies, the enfranchisement of women had to wait until the early twentieth century. The same is true of the enfranchisement of racial minorities in some countries.[14])

In light of what has been presented thus far, the main difference between the liberal state and the liberal-democratic one can also be stated as follows. Compared with the liberal state, in the liberal-democratic scheme, democracy merely means two things: the formal-legal abolition of economic-class (and later the abolition of gender, race, and ethnicity) as the qualification or eligibility for voting and being voted for, *and* granting legal-political equality to all. In this respect, the liberal-democratic notion of democracy is an improvement over the exclusionary principles of ancient Athenian democracy and Rousseau's sexist exclusion of women. However, this gain came at a great cost to the idea of democracy, and cannot compensate wholly for what was lost in the process (viz., the idea of participation and sovereignty of the people in the state).

It is worth noting at this point that in the process of transition from the liberal state into the new one, the five principles of representation went through transformations of their own as they were adapted to the realities of the new society. These changes were accompanied by an important transformation in the way the purpose and function of political parties were understood. Arguably, this latter transformation was the main vehicle of transition from the liberal state to a liberal-democratic one. The old society with its limited franchise had a small number of voters (mainly propertied classes and the middle-classes in their periphery). This meant that political parties only had to be accountable to—and represent the (class) interests of—this small minority. The expansion of the franchise in the new society greatly multiplied the number of the voters and with that, burdened the party system with what Macpherson calls the "function of blurring class lines."[15] In order to attract as many voters as possible, political parties had to "move towards a middle position . . . project an image of standing for the common good . . . [and] . . . offer a platform which . . . [was] . . . all things to all men."[16] Thus, by expansion of the franchise, the main function of the party system in the new society was transformed to that of "mediat[ing] between the demands of two classes, those with and those without substantial property."[17] The conclusion that Macpherson draws from his analysis of the purpose and function of political parties in the liberal-democratic society is that the party system as a whole has exhibited a great success in being "the means of reconciling [a] universal equal franchise with the maintenance of an unequal society."[18]

Returning to the discussion of the transformations of the five principles of representation, now that the franchise had become universalized, the principle of the legitimacy of the representative government (the "main principle") grew stronger and gained the status of a universal truth and a requirement of "civilization"—that liberal-democracy held out to the rest of the world as a model to be emulated as well as using it as an ideological weapon against its rivals. As to the "principle of collective-procedural decision-making," now that non-propertied classes were included in the franchise (and given that their interests were diametrically opposed to those of the classes which had hitherto dominated the state), the principle had to be reinterpreted. While in the old liberal state the House had to strike a balance only among the "landed interests," "manufacturing interests," "commercial interests," "creditors," and "debtors"—(in short, among the interests of the wealthy)—in legislating, now it also had to consider the interests of those whose interests potentially threatened the interests of the wealthy.[19] This new factor complicated the process of collective-procedural decision-making and elevated the importance of the skills of political maneuvering, political wheeling-dealing, and consensus building to new heights.

Apropos of the "principle of the frequent and competitive elections," as was argued earlier, the competition among the candidates for the House was often a relatively low-key local affair in the liberal state. However, once the system of universal suffrage was instituted, competition in the elections to office became wide open and more competitive, and for this reason the "principle of frequent competitive elections" acquired a new meaning and a greater importance. In the liberal-democratic society, quite to the contrary, the candidates for the House and the government offices are compelled to tailor their campaign messages and political programs in ways that would enable them to "blur"—or transcend in discourse—class divisions, and win the votes of diverse spectra of people and interests. The principle of frequent and competitive elections will be discussed in some detail later in this part and Part III.

As to the "principle of independent-virtual representation," it suffices to say that this principle played a key role in transition to the liberal-democratic state. Without the consolidation and wide acceptance of this principle, the liberal state would not have dared to open itself up to the working-class participation for fear that the actual representatives or delegates of the working-class, given that they constituted the majority, could legislate solely in their own class interests, and thus jeopardize the interests of the "whole." To make sure that this would not happen, the propertied classes had to ascertain that non-propertied classes understood that political representation was above their "short-sighted class-interests." Burke's conception of representation had this message of cessation of the class-based legislation as one of its main components.

As was also the case with the process of importing the principles of "collective-procedural decision-making" and "frequent and competitive elections," the task of politically neutralizing the non-propertied classes was both a delicate balancing act and a monumental undertaking which was accomplished successfully. In the earlier years of the liberal-democratic era, it took a great deal of complex and shrewd political maneuvering and propagandizing on the part of both the entire liberal-democratic establishment and the individual representatives to achieve what seemed to be an impossible task: to secure the support of the haves (with whose interests they were historically affiliated), and at the same time, to win the confidence and votes of the have-nots. This was achieved by "concealing" the close affiliations that tied the state to the lords of the "economic society." This came about in two ways. First, in light of the attention the new state began to pay to the general welfare of society, the affiliations in question increasingly began to appear as tolerable to many mainstream or middle-class intellectuals and critics of the new system who preferred a state dominated by the wealthy to the one that imposed the class rule of the have-nots.[20] This in turn guaranteed that the strong ties between the wealth and the state would appear less visible to the undereducated eye of the general public. Second, these affiliations became less noticeable as they grew more intricate and were blended into the general structure of the state's Keynesian intervention in the economy.[21]

Finally, as to the "principle of the distinction of the representatives," despite the eventual abolition of property qualifications for eligibility to serve as representatives, this principle continued to retain its elitist character. The representatives continued to be regarded as the individuals possessing higher moral character and higher intellectual capacity than the ordinary people. In fact, one can argue that in light of admitting non-propertied classes into the franchise, this principle acquired greater importance, especially for the members of the rapidly growing middle-classes in Europe. From the perspective of these classes, although the good thing about democracy was that it would stymie the landed aristocracy and bourgeoisie in their old practice of ramming their self-interested agenda through the legislatures (now that they were in the minority), the drawback was that the non-propertied newcomers could now do the same with their own self-interested (i.e., class-based) agenda on the sheer strength of their numbers. To prevent this, a consensus soon developed among the intellectuals of the middle-classes that the measure of regulating the admission of non-propertied classes into the franchise (e.g., setting age or minimum income requirements and excluding those who do not pay direct taxes) was not enough. In addition to such measures, the government also had to assume the responsibility of educating the ignorant and uncultured masses that were entering the franchise. But more importantly, the representative government had to resort to "mentally superior" individuals with "virtue and talent" (viz., the middle-class elites) in setting the legislative agenda and drafting bills. To counteract the negative effects of extending the franchise to the "untutored masses," it was deemed necessary that the power of drafting bills be concentrated in the hands of the "instructed minority." In fact, this view was articulated forcefully by J. S. Mill in England who led a revisionist humanist trend in liberalism.[22]

*   *   *   *   *

The earlier claim that the liberalization of democracy was achieved at the cost of perverting the original meaning of the idea of democracy is vividly illustrated in J. S. Mill's *Considerations on Representative Government*. Although the work starts as a document in strong support of direct-participatory democracy (especially Chapters II and III), it quickly degenerates into prescribing a remarkably elitist conception of government that not only disqualifies ordinary people from partaking in decision-making, but also argues for a system of political inequalities.[23] Chapter II develops two main criteria for a good form of government: the ability to improve the character of the people (individually and collectively, on the one hand, and morally as well as intellectually, on the other hand) and the ability to utilize the community's virtue and character.[24] Chapter III affirms that "ideally the best form of government is that in which the sovereignty . . . is vested in the entire aggregate of the community, [and in] every citizen."[25] After presenting ancient Athens' experience with direct democracy in a glowingly positive light, the chapter also declares that the ultimate end of a good form of government (and thus the function of the ideally best form of government) should be to provide the maximum participation for all members in the sovereignty.[26] *However*, Mill takes the reader completely by surprise in the very last sentence of Chapter III when he suddenly and hastily argues that since direct-participatory democracy is impossible in a large nation-state, "the ideal type of a perfect government must be representative."[27] This sets the theme for the rest of the book that goes on to advance an utterly elitist conception of government which he believes to embody the "true" meaning of democracy.[28]

In his conception of representative government, Mill is strictly a Burkean. This in part is manifest in his view that deliberation is the *main* function of the representative assembly.[29] "[T]alking and discussion are their [representatives'] proper business."[30] The tasks of administrating (i.e., "doing") in Mill's conception are to be given to a highly skilled class of administrators (the cabinet) who would be selected by the assembly.[31] The more important function of legislating the laws, on the other hand, would be the exclusive prerogative of a special commission of experts (morally-intellectually superb and disinterested individuals) who would be appointed by the Crown. The role of the representatives in legislative matters would be limited to approving, rejecting, or recommending "reconsideration or improvement" to the laws framed by this special commission. Broadly speaking, the main function of the representative assembly consists in "controlling the business of the government" by being "the office of control and criticism," i.e., by watching "the actual conduct of affairs" by the administration.[32]

What is blatantly undemocratic and truly disturbing in Mill's theory of representative government is its attitude toward the ordinary (working) people. Mill regards the laboring-classes as "mere masses of ignorance, stupidity, and baleful prejudice."[33] The prospect of the rule by majority (the masses), and consequently, the prospect of the oppression of the minority (the upper- and middle-classes), as well as the idea of political equality frightened Mill tremendously. For Mill, the notions of the rule by majority and political equality represented a "false democracy" or "collective despotism," which he believed was the order of the day in the United States of his time.[34] To ensure that a false democracy does not prevail in England, Mill supported the utterly inegalitarian idea of "plural voting."[35] In the same vein, to guard against the oppression of the minorities (the upper- and middle-classes), he also supported "proportional representation."[36]

In reading *Considerations on Representative Government*, one develops the sense that Mill's negative characterizations of ordinary (working) citizens are not merely prejudicial or empirical-factual in nature, but rather stem from his deep-seated philosophical convictions on the nature of labor, society, and civilization in general.[37] Nevertheless, on the strength of the earlier chapters and other works of his, one is inclined to assume that Mill probably believed that in the future the ordinary (working) people could be raised to higher levels of moral or intellectual competence, in which case they should be entitled to political equality and thus to a full participation in the franchise. Doing so could explain away the inegalitarian strands in his thought and persuade one to believe that Mill thought keeping the ordinary people away from full participation was a temporary cautionary measure. Yet, one finds it hard to shake off the suspicion that the functions of educating the untutored masses that Mill conferred upon the government were not intended to raise the laboring classes to the same levels of intelligence and moral capacity achieved by the members of the upper- and middle-classes, but primarily to inculcate in them the culture of reverence and deference for the wiser and learned members of the society. In other words, the main idea for Mill, it seems, was to educate the laborers enough so that they would come to understand how intellectually and morally incapable they were; and thus, they would realize that they should not "waste" their votes on the "likes of themselves" and instead vote for "their employers and their social and intellectual superiors."[38] By many indications, it seems that Mill only had the members of the middle- and upper-classes in mind whenever he spoke of the idea of citizens partaking in governing in the earlier chapters of the work.[39] Finally, the other undemocratic strand in Mill's theory is that, by the idea of utilizing the talent and virtue of the community in governing, Mill did not have in mind the idea of drawing on the wisdom of the whole community or the collective wisdom of the people, but merely the electing and recruiting of the "wisest members," the elite, in the community into the government (as representatives and legislatures).[40]

It should be noted here that in his characterization of American democracy as "false democracy," as well as in his approach to the question of combining democracy with liberalism, J. S. Mill was influenced by Tocqueville's *Democracy in America*. In this work, Tocqueville had admired and idealized the early-to-mid nineteenth century practice of participatory democracy in the American New England states, especially in Connecticut and Massachusetts.[41] J. S. Mill's ambivalence on the question of the participation of ordinary people, one would suspect, mirrored the ambivalence that pervaded some parts of *Democracy in America*.[42] Although Tocqueville was impressed by the positive aspects of democracy in the United States (i.e., its power to energize the people and contribute to their civic education), he believed that democracy could promote extreme "individualism" and materialism, and thus threaten liberties by giving rise to a new kind of despotism (the pressure of public opinion and hence the "tyranny of majority").[43] Moreover, based on his observations, he believed that democracy was not a particularly good system of decision-making, for it had a tendency to undervalue the importance of expert knowledge and attract those who were "inferior in capacity and morality" to public service.[44] Tocqueville's observations, in fact, contradicted Madison's argument that the representatives would turn out to be the men of "wisdom" and "virtue."[45]

*   *   *   *   *

In closing the discussion of the larger question of the rise of the liberal-democratic state, it should also be added that the principles of representative government were not all that was imported into liberal democracy from the liberal state. Liberal democracy also inherited the totality of beliefs and assumptions on Man and society that constituted the core of liberalism. The most important of these beliefs and assumptions are often listed as commitments to individualism, individual freedoms, the constitutional protection of civil liberties and rights, the moral and legal-political equality of citizens, the separation of private and public spheres, the belief in the sovereignty and independence of the individual in her private sphere of life, the moral and intellectual neutrality of the state, and the belief in the separation of the state from society and economy. What changed in this bequeathal was that the status of citizenship was extended to all adults. Finally, it should be mentioned that the adoption of the democratic franchise by the liberal state was not entirely a sinister political design. It did not come about only as a counter-measure intended to halt the working-class revolutions, save the liberal state, and thus legitimize the system of substitutive inequalities. The movement toward the adoption of liberal democracy had also a genuinely moral component that was championed by the humanist-minded intellectuals of the nineteenth century such as J. S. Mill and T. H. Green. In the case of J. S. Mill, despite the fact that he was not a "full egalitarian" in his attitude toward the working-classes, his conception of representative government, as Macpherson has argued, had a "moral dimension" that saw democracy "primarily as a means of individual self-development."[46]

Notes

1. Defined broadly, "economic society" here denotes the domain of economic activities, institutions, and entities that include, among others, the relations governing production and distribution, "means of production," markets, financial institutions, economic firms and corporations, and the activity of work itself. The "economic society" is distinguished from the "civil society" and the "political society." The term "civil society" de-

notes the sum total of all societal entities and spheres of activities that deal with, or relate to, the question of life in the society. Simply put, civil society is the sphere of life outside of the workplace, as the "economic society" is the sphere of life in the workplace, and in this sense, it is separate and distinct from the latter society. The "civil society" is also apart and distinct from the state (the "political society"). These three domains are defined and discussed in detail in Chapter 4 of *Direct-Deliberative e-Democracy*.

    2. Macpherson (1965), p.8.

    3. Ibid., pp.8-9, italics added.

    4. Ibid. p.9.

    5. Ibid., pp.9-10.

    6. One should add that the general model of transition from the liberal state to the liberal-democratic one depicted by Macpherson in the passage quoted above applies mainly to the case of Europe, particularly to England. This model does not fit the U.S. situation in a straightforward way. The "liberal" Constitution of 1787 in the U.S. did not succeed in transforming the new country into a "liberal society" immediately. This was mainly due to the fact that the new constitution could not erode the localist and participatory democratic tradition that had taken hold in America in the seventeenth and eighteenth centuries. Although the "aristocratically" inclined and predominantly capitalist (and/or capitalistic-minded) Federalists won the battle of the constitution, it was mostly the Anti-Federalists' conception of representation and politics that prevailed in practice in the early republic. Thus, the white Americans kept their "conservative" democratic tradition, and used the new constitution in ways that helped them (primarily the middle-class entrepreneurs, farmers, and artisans) expand and prosper. (This was primarily the Jacksonian America that Tocqueville witnessed.) Viewed from this angle, the "bourgeois" revolution of 1776 worked more to the benefit of the middle-class Anti-Federalists than in accordance with the original wishes of the Federalist "moneyed men." The American finance and industrial capitalism that came to assert its power after the Civil War grew primarily out of the Anti-Federalist or Republican type of entrepreneurs who managed to accumulate and expand. (See Wood (1997), especially p.147. Wood argues that "hard-working Republican 'laborers' were the main force behind America's capitalist market revolution. For good or for ill, American capitalism was created by American democracy." See also ibid., p.147n37.) *The "liberal state" (and the "liberal society") eventually managed to establish itself in the United States only in the post-Civil War era.* And even then, it resembled only some aspects of the classical case of the liberal state (e.g., England of 1814-1914, which is described as the "golden age of liberal theory and practice" by Gray (1985), p.27). This was primarily the 1865-1900 period in American history. (See Beaud (2001) and Duboff (1989) on the expansion of American capitalism in this era.) It was in this era that the Federalists' original aim of having the national government assert its supremacy over the state governments was finally realized. The victorious industrial and finance capitalism in the post-Civil War period began to expand faster than it had ever done in the past. It consolidated its powers in corporations and trusts and instrumentalized the federal and local governments in ways that left the "moneyed classes" in charge of running the free-markets in very much the same ways that their counterparts did in England for much of the nineteenth-century. Constitutional seperation of powers, the autonomy of the local and state governments, and the system of checks and balances enabled the new rising propertied and moneyed classes to prevent governments from intervening in economy. This allowed the wealthy to exercise their own "sovereignty" over the markets. Baran and Sweezy's description of the situation is vivid and thorough:

before the end of the nineteenth-century . . . [t]he United States became a sort of utopia for the private sovereignties of property and business. The very structure of government prevented effective action in many areas of the economy or social life. . . . And even where this was not so, the system of political representation, together with the absence of responsible political parties, gave an effective veto power to temporary or permanent coalitions of vested interests. The positive power of government has tended to be narrowly confined to a few functions which could command the approval of substantially all elements of the moneyed classes: extending the national territory and protecting the interests of American businessmen and investors abroad, activities which throughout the nation's history have been the first concern of the federal government; perfecting and protecting property rights at home, carving up the public domain among the most powerful and insistent claimants; providing a minimum infrastructure for the profitable operation of private business; passing out favors and subsidies in accordance with the well-known principles of the logroll and the pork barrel. Until the New Deal period of the 1930's, there was not even any pretense that promoting the welfare of the lower classes was a responsibility of government. . . .

This was the situation which prevailed at the time of the collapse of the boom of the 1920's (Baran and Sweezy 1968, pp.158-57).

The late nineteenth century was also the era of workers' strikes and unrests—the railroad strikes of 1877 being the most important event. (See Zinn (1980), Chapters 10-12, for a vivid image of the labor strikes in this period.) Though the Constitutional Amendment XV in 1870 enfranchised all males, during this period, in practice, millions of native Americans, American-born Asians, Latin-American and European immigrant workers, and blacks in the South were kept out of the franchise. Then came the early twentieth century with its further intensification of workers' struggles (the "Socialist Movement") and the women's struggle for voting rights. And by the time the Great Depression appeared on the American horizon, the country was ready for a great change: *a transition from the liberal state-society to the "liberal-democratic" one* that was led by Franklin D. Roosevelt's New Deal. This point will be revisited later in this part of the volume and in Part III.

7. Macpherson (1965), p.10.

8. The term "swinish multitude" is a reference to Edmund Burke's characterization of the ordinary people. As was stated in the preceding chapters, Burke's conception of virtual-independent representation had a shade of class compromise and cessation of class-warfare, which proved handy in transition from the liberal state to the liberal-democratic one.

9. For instance, James Mill advocated the exclusion from franchise of all men and women under the age of 40, as well as the exclusion of the "poorest one-third of the males over 40" (Macpherson 1977, p.38). (See also Birch for his discussion of the position of James Mill (Birch 1971, p.55 and pp.60-62.) The scheme of voting that was widely discussed was the "*plural voting*," proposed by Thomas Hare and supported by J. S. Mill. In the "plural voting" scheme, educated individuals (primarily the members of

the upper- and middle-classes) were entitled to have more votes, up to five or more, than the uneducated ones (the members of the lower classes), who were entitled to only one vote per person.

10. The reform acts of 1832, 1867, 1884, and 1885 extended voting rights in England. The reform act of 1832 defined the eligible voter as any male who owned a household worth of 10 British Pound. This act enfranchised the middle class and as a result, 20% of the male population were qualified to vote. The reform act of 1867 enlarged the pool of voters by allowing many workers to vote. The voting acts of 1884 and 1885 added the agricultural workers to the pool, thus breaking the class barrier completely. The age barrier for men was broken in 1919 when all men over 21 were enfranchised. The gender barrier was cracked in 1918 when women over the age of 30 were enfranchised, and it finally came down in 1928 by the Equal Franchise Act which reduced the voting age for women to 21. (Source: Eric Hobsbawm's *Industry and Empire: The Birth of the Industrial Revolution*.) One should note that the progression in the expansion of democratic franchise present in these acts closely parallel the progression in the position of Jeremy Bentham's thought on the question expressed in the period of 1791-1817. (See Macpherson 1977, p.35, and p.37.)

11. Macpherson (1977), p.35.

12. Woodrow Wilson made the statement in his address to the U.S. Congress on April 2, 1917. The paraphrases of "good thing," "bad thing," and "full acceptance into the ranks of respectability" are taken from Macpherson (1965), pp.1-2. See Chapter 2, including notes 48 and 56, for some comments on the negative use of the word "democracy" in the United States and Europe in the pre-World War I period.

13. Macpherson (1965), p.5.

14. In England, the full enfranchisement of the women had to wait until 1928. In the case of the United States, women were admitted into the franchise in 1920, and blacks in the South in 1964, when the poll tax was dropped as a qualification for voting eligibility.

15. Macpherson (1977), p.66.

16. Ibid.

17. Macpherson (1977), p.68. (Also see Macpherson 1973, p.191.)

18. Ibid., p.69. Macpherson further argues that this success has come at the cost of making the parties "less responsible to the electorate" (ibid., p.67) and also by "blurring the issues and by diminishing government's responsibility to electorates" (ibid., p.69). Moreover, he argues that in appealing to a mass electorate, the internal structures of parties became less democratic for, "[e]ffective organization required centrally controlled party machines. . . . The central party leadership was therefore able to control its M.P.s [in the case of England]. The main power fell to the party leaders in Parliament" (ibid., p.68).

19. Madison's own phrases (Pitkin 1967, p.192).

20. The role played by middle-class intellectuals will be discussed further in what follows.

21. This was done in part by allowing the chambers of commerce and influential members of the business world to draft or influence the state's economic policies, and in part by instituting a system of privileges and subsidies, as well as a system of immunities and bailouts for big corporations and their wealthy owners.

22. That J. S. Mill's views on government were completely committed to advancing the middle-class causes is forcefully argued by Currin Shields in his editorial comments on Mill's *Considerations on Representative Government* (1958), pp.xxxvii-xl.

23. Unfortunately, those who treat J. S. Mill as the champion of participatory democracy often fail to take into account this negative side of Mill.

24. Mill (1958 [1861]), pp.25-28.

25. Ibid., p.42.

26. Ibid., p.55. J. S. Mill's position that participation was a good thing in itself was an improvement over his father's and earlier Utilitarians' who approved of participation (voting and representation) mainly because they believed it would help to maximize private benefits (Birch 1971, p.66).

27. Ibid., p.55.

28. As will be seen later, J. S. Mill believed that his conception of representative government sidestepped the problems of "false democracy" which he believed reigned in the United States of his time, e.g., Mill (1958 [1861]), p.114.

29. Especially in his Chapter V. Mill was also a Burkean in the sense that he believed the parliament was the assembly of one nation with one interest. In practice, according to Thompson, during his three years of service as an M.P. from Westminster, Mill "behaved himself much like a Burkean representative" (Thompson 1976, p.116). It should be noted that, as Pitkin argues, Mill's view on representation has some Madisonian elements in it in that he defines classes in terms of factions and believes that "the selfish interests [represented by these classes in the legislature] will cancel each other out" (Pitkin 1967, pp.202-03). The main difference between Madison and Mill's views is that for Mill, "not only that the selfish interests balance, but also that there be a minority who act on the basis of 'reason, justice, and the good of the whole'" (ibid., p.203).

30. Mill (1958 [1861]), p.83.

31. Ibid., p.83.

32. Ibid., p.70 and p.84.

33. Ibid., p.25.

34. Ibid., p.114 and p.119.

35. Mill (1958 [1861]), Chapter VIII. As was mentioned in note 9 above, in *plural voting*, the educated individuals (primarily the members of the upper- and middle-classes) were entitled to have more votes, up to five or more, than the uneducated ones (the members of the lower classes) who were entitled to only one vote per person. Mill bluntly states that plural voting is intended to guard against the "class legislation of the uneducated" and "to prevent the laboring class from becoming preponderant in Parliament" (ibid., p.139 and p.141, respectively).

36. In Chapter VII. The proportional representation scheme proposed by Thomas Hare used a national constituency rather than a local constituency. In Mill's view, this scheme not only would assure that minorities (the upper- and middle-classes) were represented in the parliament, but also it would elect "the very *elite* of the country" (Mill 1958[1861], p.113, original italics). Moreover, given that these minorities would get to elect the best minds using Hare's method, "[m]ajorities [the lower classes] would be compelled to look out for members of a much higher caliber" (ibid.).

37. By many indications, these convictions are class based. For Mill, the negative characterizations of the masses did not stem from the fact that they had not had the opportunity to educate themselves, but rather resulted from their socio-economic roles in society as *manual laborers*. The following passage illustrates this point: "how little there is in most men's ordinary life to give any largeness either to their conceptions or to their sentiments. Their work is a routine . . . neither the thing done nor the process of doing it introduces the mind to thoughts or feelings extending beyond individuals; if instructive books are within their reach, there is no stimulus to read them" (Mill 1958[1861], p.53).

One is justified in assuming that Mill believed that the problem of the moral and intellectual incompetence of the laboring classes would persist so long as the capitalist mode of production (e.g., division of labor) prevailed. And Mill was not an anti-capitalist thinker. As Macpherson has shown, Mill regarded the capitalist mode of production as just, albeit he believed that its mode of distribution was unjust and had to be made more egalitarian (Macpherson 1977, pp.50-74). A more egalitarian distribution, one can argue, could only mitigate the problem of injustice and not eradicate it. It then follows that for Mill the morally and intellectually inferior status of the laboring masses was considered a permanent state of affairs. Moreover, Mill regarded "realized property," "knowledge," and "individuality" as main features of a civilized society, all of which were absent in the non-propertied classes.

38. This is how Smart seems to interpret Mill's claim that the "electors should choose as their representatives wiser men than themselves" (Smart 1991, p.113).

39. This negative portrayal of Mill is based exclusively on his *Considerations on Representative Government* (1958[1861]). In contrast to this work, the younger Mill in his *Principles of Political Economy* (1848) sounds optimistic and gives a sense that he believes the members of the laboring classes could eventually be elevated (intellectually and morally) to the levels of the members of the upper- and middle-classes. It is based on this positive reading of the *younger* Mill that Macpherson classifies Mill's model of government as a "developmental model" of democracy (Macpherson 1977). Moreover, it is also based on this positive reading that Pateman (1970) and other participatory democrats regard Mill as a champion of participatory democracy.

40. Mill (1958[1861]), p.27. According to Held, the elites in Mill's system could be regarded as "a modern version of philosopher-kings" (Held 1987, p.101). Shields is utterly convinced that for Mill, "the people are supreme in theory but in practice are permitted to play no important role in exercising authority" (in his editorial comments in Mill 1958[1861], p.xxv). "By the time Mill began his major political treatise," according to Shields, "he had abandoned the principles of democratic rule" (ibid., p.xxiv). Shields goes on to quote Mill to the effect that he had "grave fears about the leveling influence of democracy on civilization" (ibid., note 6).

41. According to Tocqueville, in these states, "the sum total of men's education is directed toward politics," unlike Europe where the main goal of education was to prepare the citizens for private life (Tocqueville 2000, Vol.1, pp.291-92). Although the American citizen "is made familiar with the history of his native country and the principal features of the constitution that governs it," as Tocqueville believed, the citizen's genuine education came "principally from experience" (ibid., p.289, and p.291).

42. Tocqueville's ambivalence about democracy is vividly present in Vol.1, Part II, Chapters 5 and 6, pp.219-22 of Tocqueville (2000) where he compares democracy with aristocracy. (see also pp.234-35.)

43. Tocqueville believed that in the United States, democracy entailed not just political equality (representative government and extended franchise), but also social equality. The social effects of the latter were to give rise to what he called "individualism." One major aspect of individualism was the belief that all ideas must be put to the test of individual reason and that all individuals had the capacity to do so, and consequently, the opinions of all had equal status. Another consequence of this attitude was the devaluation of expert opinion and intellectual authority. The second major aspect of individualism, on the other hand, was egoism and the prevalence of self-interested pursuits. Tocqueville argued that the first aspect could lead to conformity—(if all opinions counted equally, this would mean that the decisions made by the majority should be right, which would

imply that the minority was wrong and thus they should give up their opinion and conform to what has been decided by the majority, hence the pressure of public opinion and the tyranny of the majority). The second aspect, on the other hand, would lead to apathy toward public concerns and thus to leaving the business of the public and governing to politicians. Thus, both aspects would end up threatening individual liberties. Contrary to the elitist solutions proposed by Mill to problems posed by democracy, Tocqueville took an optimistic view of things and believed that Americans had already found ways of mitigating the destructive tendencies of these problems. Tocqueville believed that the Americans' "free *moeurs*" and their participation in local governments, as well as their conception of "self-interest well understood," their "decentralized administration," extensive system of free associations, and free press were counteracting the tendency of democracy toward tyranny. It is important to note, in light of the fact that Tocqueville was coming from a continent of sharp class divides and unrest, especially his native country—(also given that he was infatuated by the material success and wealth of the Americans)—his eyes were not trained to see somewhat subtler class divisions and inequalities that prevailed in the U.S. at the time. Tocqueville's assertion that there existed a "general equality of conditions among the people," according to historian Howard Zinn, "was not in accord with the facts" (Zinn 1980, p.213). (Zinn declares this on the authority of *Jacksonian America* (1969) by Edward Pessen, a historian of Jacksonian society.)

44. Tocqueville (2000), Vol.1 p.223. According to Tocqueville, "[w]hile the natural instincts of democracy bring the people to keep distinguished men away from power, an instinct no less strong brings the latter to distance themselves from a political career" (Tocqueville 2000, Vol.1, p.189). For Tocqueville, the laws made in democracy were "almost always defective or unreasonable" (ibid., p.222). However, Tocqueville held the view that democracy was the "most appropriate of all [governments]" insofar as material prosperity is concerned (ibid.). (See also pp.234-35.) Following in the footsteps of Tocqueville, Mill claimed that in America, "the first minds in the country [are] being as effectually shut out from the national representation, and from public functions generally, as if they were under a formal disqualification" (Mill 1958[1861], p.129, also p.115).

45. Upon his visit to the House of Representatives, Tocqueville found its members to be "vulgar elements" and "obscure persons, whose name[s] furnish no image to one's thought," and as being "for the most part, village attorneys, those in trade, or even belonging to the lowest class," who "do not always know how to write correctly" (Tocqueville 2000, p.191). In contrast, Tocqueville found the members of the Senate to be the individuals of caliber idealized by Madison, i.e., as men who had the "monopoly on talents and enlightenment" (ibid., pp.191-92). However, the problem was that these individuals were the elites of their respective states in the Union and were elected by the assemblies of their states (and not in federal elections)—and thus could not be regarded as resembling the representatives idealized by Madison's conception of representatives.

46. Macpherson (1997), p.22. The "developmental model" in Macpherson's classification of the models of democracy was preceded by the "protective model" of Bentham and James Mill, which was more realistic and closely resembled how the franchise was developed in England. Although the basic idea of the protective model was to protect the citizen—i.e., the property-owning male citizen—against governmental intrusions, the model was also concerned with protecting the position of the rich as the politically dominant class. Macpherson quotes James Mill to the effect that the "business of the government is properly the business of the rich" (Macpherson 1977, p.42). Macpherson characterizes the "spirit" of this model by the following statement: "the democratic franchise would not only protect the citizens, but would even improve the performance of the rich

as governors. It is scarcely a spirit of equality" (ibid.). The characterization of Mill as not being a "full egalitarian" is taken from Macpherson (ibid., p.59).

Chapter 9

Schumpeter and the Liberal-Democratic Conception of Democracy

The rise and consolidation of the liberal-democratic state coincided with another development that paralleled it in the realm of theory, viz., the revision of liberalism. This development, which took place during the late nineteenth century and early twentieth century, contributed immensely to the rise and consolidation of the new state, and at the same time was affected by this process in profound ways. As democracy became liberalized, liberalism became "democratized"—or rather one should say, it became socialized.[1] In this process, liberalism gradually gave up its fixation on market-based notions of freedom and the anti-regulatory postures of its classical period and moved slowly toward embracing views that were sympathetic to the social-developmental policies of the welfare state. This development took place mainly as a consequence of the loss of faith of the proponents of liberalism in the ability of the market to regulate itself (or head off economic crises for that matter), and also as a consequence of the success the state exhibited in its intervention in the market in an attempt to save it. Another factor that contributed to this development was the state's successful management of the wartime economy during the two world wars. In England, the socialization of liberalism began with J. S. Mill on philosophical grounds in the mid-1840s. With a new attitude toward trade unions and socialist experimentation, Mill broke from the tradition of classical liberalism. This heresy was followed by T. H. Green and Bernard Bosanquet who argued for a more positive conception of freedom, thus arguing for an activist conception of government. The most prominent figure of the new revisionist liberalism in the England of the early twentieth century was L. T. Hobhouse whose conception of liberalism combined J. S. Mill's and T. H. Green's. [2]

In the case of the United States, the social legislations that began after the economic crash of 1929 (the Social Security Act of 1935 being its crown jewel) followed a course that culminated in the Great Society reforms of the Lyndon Johnson administration in the mid-1960s.[3] In the realm of theory, no longer were liberty and free-markets (or property rights) assumed to mean the same thing or serve as a co-condition for one another, as was the case with classical liberalism. No longer was the government deemed to be the main source of evil in society. (The defense of this latter view, as well as the arguments in its support, became mainly, if not exclusively, a property of those liberals who refused to acknowledge that the world had changed since the time of Adam Smith—the present day neo-conservatives and libertarians.) In the United States, the last bastion of classical liberalism, the socialization of liberalism went so far that, since the late 1960s, liberalism became synonymous with welfare-statism in U.S. politics. The most important theoretical development in this period in the United States was the publication of John Rawls' *A Theory of Justice* in 1971, which put forth an egalitarian conception of liberalism by proposing two principles of social justice without compromising the priority of liberty. However, with the rise of the tide of the shift to the right that swept Ronald Reagan into office in 1981, and along with it the rhetorical and legislative assaults on the welfare-regulatory state, as well as the mass of legislations directed at deregulating the U.S. economy that ensued (sometimes dubbed as the "neoclassical" restructuring of the economy), the term "liberal" took on a derogatory connotation that implied a proponent of big government, and "a big welfare spender." This connotation in American politics continued into the twenty-first century.[4]

Having presented this brief historical account of how the idea of democracy became liberalized (or how liberalism became democratized), the questions to be addressed now are as follows. What theoretical shape did the idea of democracy assume after it was liberalized (or after liberalism became democratized)? And given that the original meaning of democracy was abandoned, what came to constitute the meaning or content of democracy? Moreover, how were "the main principle" of representative government and its four accompanying principles affected or altered in the process of importation from the liberal state into the liberal-democratic one?[5]

The answer to these questions must be sought first and foremost in Joseph A. Schumpeter's seminal work *Capitalism, Socialism, and Democracy* which was published in 1942. This work, as will be seen below, posited democracy not as a set of moral ends or a form of social organization, but exclusively as a "method" or a *technique* for selecting the individuals who would be endowed with the power to rule the people. *Capitalism, Socialism, and Democracy* set the tone for what was to follow. As will be shown shortly, the subsequent developments in the theory of democracy in the twentieth century followed the lead of this work. Finally, before starting to examine this work, it is worth noting that, although Schumpeter is mainly known as a prominent economist, he had a passion for politics, and in fact had acquired some experience in working in government in Europe before immigrating to the United States in 1932.[6] As will be seen, the fact that by profession and training he was an economist influenced his approach to the question of democracy, as it conditioned his thinking to see strong resemblances between politics, on the one hand, and the operations of the market economics, on the other hand.

Now, of the six principles of liberal democracy directly related to the question of governing that were discussed in Chapters 7 and 8—(the new principle of "universal suffrage," the "main principle of representative government," and its four accompanying principles: the "principle of collective-procedural decision-making," the "principle of the distinction of representatives," the "principle of independent-virtual representatives," and the "principle of frequent and competitive elections")—Schumpeter had the least esteem for the first and the most enthusiasm for the last three. To begin with, Schumpeter ventured into arguing that the idea of limiting the domain of universal suffrage was not completely incompatible with democracy, or as he put it, "discrimination can never be entirely absent."[7] Schumpeter does not appear to be wholly unsympathetic to some arguments in favor of using property qualifications (or as he puts it, "one's ability to support oneself"), race, gender, and religious affiliation as criteria for excluding some members of society from universal suffrage.[8] Regarding his high esteem for the last three principles, as will be seen, Schumpeter was a strong believer in the principle of independent-virtual representation in its Burkean sense. As he put it once, this idea was "a principle that has indeed been universally recognized by constitutions and political theory ever since Edmund Burke's time."[9] He saw representation as an important component of the division of labor in governing: the people would elect the representatives and then stand back and let their representatives exercise their powers of judgement and decision-making. Schumpeter was a staunch supporter of the idea of keeping ordinary people away from the business of governing. As to the principle of the distinction of the representatives, Schumpeter's theory assigned an important role to high-quality leadership. As a matter of fact, it will be seen later that he was in favor of socially breeding a special class of individuals for leadership positions. Finally, as to the importance of the principle of competitive elections for democracy, Schumpeter took it to be the hallmark of the idea of democracy. His enthusiasm for this principle is perhaps the main reason why he referred to his theory of democracy as "the theory of competitive leadership."[10]

Before presenting Schumpeter's theory of democracy, it would be helpful first to give an outline of what he refers to as "the classical doctrine of democracy," in opposition to which he presents his own theory. According to Schumpeter, the classical doctrine, to begin with, was ideological (or an "ideology") and "religious" in nature.[11] It rose in the seventeenth and eighteenth century as, what Schumpeter calls, a "substitute" to fill the vacuum created by the loss of faith in the higher authority of God, which had disappeared from the face of the earth at the time. Moreover, according to Schumpeter, this ideology rested on "a rationalist scheme of human action and of the values of life."[12] And for these reasons, it had no relevance to reality, and was only applicable to "small and primitive societies."[13] By some indications, it seems that Schumpeter's understanding of the historical and theoretical roots of the "classical doctrine" is somewhat confused. On the one hand, he locates the formulation of the classical doctrine in the eighteenth century and refers to it as "the eighteenth-century philosophy of democracy."[14] As a result, the reader's attention, naturally, is directed to Rousseau—he mentions Rousseau by name only once. On the other hand, he keeps referring to the "classical doctrine" as a "utilitarian" theory, in which case the attention of the reader is diverted to Bentham and J. S. Mill in the late eighteenth and early nineteenth centuries whose names he mentions once or twice. [15] Despite this confusion, Schumpeter's presentation of what he takes to be the "classical doctrine of democracy" is clear and well-defined. According to Schumpeter, democracy in the classical doctrine is:

> [an] institutional arrangement for arriving at political decisions which realizes the common good by making the people itself decide issues through the election of individuals [actual representatives or delegates] who are to assemble in order to carry out its will. [16]

Alternately stated, democracy in this doctrine is:

> the proposition that "the people" hold a definite rational opinion about every individual question and that they give effect to this opinion . . . by choosing [actual] "representatives"[i.e., delegates] who will see to it that that opinion is carried out. Thus the selection of the representatives is made secondary to the primary purpose of the democratic arrangement which is to vest the power of deciding political issues in the electorate. [17]

In other words, in the "classical doctrine," democracy is conceptualized as a system of social-political institutions put in place for the purpose of realizing a set of moral ends (the common good). In this conception, the sovereignty of the people is the primary feature of democracy, and the election of the representatives to exert this sovereignty is a secondary concern. Based on what Schumpeter provides in the way of defining the classical doctrine, especially given that the representatives function as delegates, it seems more fitting to characterize the classical doctrine as a doctrine of "delegatory democracy." The need for delegation, according to Schumpeter, arises from the fact that it is "highly inconvenient" to have all citizens attend the business of ruling. [18] Citizens' direct participation would only be needed for "the most important decisions," i.e., for the referenda. [19] The rest of the time, the people would only need to instruct their delegates or deputies on how to proceed when all the delegates assemble to make legislative decisions. The delegates in the assembly then "will voice, reflect, or represent the will of the electorate." [20]

Now, insofar as the procedural or structural aspects of decision-making are concerned, Schumpeter's "classical doctrine" does not seem to have much in common with Rousseau's, for his requires citizens' *direct* participation. In Rousseau's conception, it is the people themselves, and not their deputies, who make the decisions—albeit, as was discussed in Chapter 3, Rousseau would also go along with a delegatory form of democracy, provided that it is premised on the exercise of the people's authority to "ratify" "*in person*" or reject the laws legislated by their delegates. [21] As a matter of fact, Schumpeter's "classical doctrine," save for the referenda, has more in common with the Soviet model of democracy or with the model of the Paris Commune than with Rousseau's or Bentham's model of democracy.[22]

In order to demolish the classical conception of democracy, Schumpeter first breaks it down to a set of simpler propositions (or "implications" as he calls them) that taken together would constitute the conception. He then goes on to show the inadequacies of these propositions, be it their logical or theoretical inconsistencies or their unrealistic assumptions. What follows is an outline of how Schumpeter breaks down the classical doctrine.

A. The classical doctrine assumes that "there exists" a thing called "Common Good . . . which is *always simple to define*." It further assumes that this common good would be understandable by "*every* normal person . . . by means of rational argument." (Lack of this understanding is taken as a sign of "stupidity and anti-social interest.")

B. This common good implies that there exist "*definite* answers to all questions so that *every* social fact and *every* measure taken or to be taken can *unequivocally* be classed as '*good*' or '*bad.*'"

C. There also exists a "Common Will of the people (=will of *all* reasonable individuals) that is *exactly* coterminous with the common good or interest or welfare or happiness."

D. Discounting "stupidity and sinister interests," there would be no "disagreement and account for the presence of an opposition." What then would exist is "a difference of opinion as to the speed with which the goal, itself common to nearly all, is to be approached."

E. Therefore, "*every* member of the community, conscious of that goal, knowing his or her mind, discerning what is good and what is bad, takes part, actively and responsibly, in furthering the former [the good] and fighting the latter [the bad] and *all* the members taken together control their public affairs."[23]

Before discussing how Schumpeter goes about rejecting these presuppositions, it should be mentioned that, even though "the classical doctrine" does not fit Rousseau's conception procedurally or structurally, in light of this outline, it does come close to Rousseau's insofar as some of its philosophical assumptions about the existence of the common good and the faith in the rational-moral competence of ordinary people are concerned. The main difference is that the Rousseauean notion of the general will is not the same as the "Common Will" or "will of all" in item C above. As a matter of fact, Rousseau was against the rule by the "will of all" which he took to mean the sum of the private wills. As to rejecting these presuppositions, Schumpeter argues that

F. The common good is a myth: "[t]here is . . . no such thing as a uniquely determined common good that all people could agree on or be made to agree on by the force of rational argument . . . to different individuals and groups the common good is bound to mean different things."[24]

G. Granting that "a sufficiently definite common good . . . [is] proved acceptable to *all*, this would not imply equally definite answers to *individual* issues."[25]

H. Now, "as a consequence" of these two propositions, "the particular concept of the will of the people or the *volonte' generale* that the utilitarians made their own vanishes into thin air." The idea of the "will of all" is "derived" from the "wills of the individuals" by the exponents of the classical doctrine, i.e., the "utilitarians." "And unless there is a center, the common good, toward which, in the long run at least, *all* individual wills gravitate, we shall not get that particular type of 'natural' *volonte' generale.*" The *volonte' generale* can only be constructed artificially by means of rational discussions by making use of "utilitarian reason." Thus not only *volonte' generale* does not exist, but also its "ethical dignity" is called into question, for it is assumed to be manufactured and then attributed to the people.[26]

I. Besides, there do not exist such things as the wills of the individuals to be used to form the *volonte' generale* from, because the individual citizens do not know exactly what they want. They do not have "definite" and "independent" notions of what they want.[27]

J. Moreover, the individual citizens do not have definite wills because they lack the intellectual-rational capacities to form them.[28]

K. Furthermore, the wills of the citizens cannot be "respected," valued, or trusted, for the citizens are morally corrupt.[29]

L. Also, the individual citizens are not interested in participating in politics.[30]

M. In addition, given items J through L, the individual citizens are susceptible to falling for demagogues.[31]

N. Moreover, even if we assume that the citizens have "definite" and "independent" wills, as well as having the intellectual-rational capacities (and moral competence), "it would not necessarily follow that the political decisions produced by that process from the raw material of those individual volitions would represent anything that could in any convincing sense be called the will of the people. It is not only conceivable but, whenever individual wills are much divided, very likely that the political decisions produced will not conform to 'what people really want'. Nor can it be replied that, if not exactly what they want, they will get a 'fair compromise'."[32]

O. Finally, given that the differences of opinion among the people would exist, albeit they would be "difference[s] of opinion as to the speed with which the goal, itself common to nearly all, is to be approached"(proposition D above), granting that then the decisions must be made by a simple majority vote, then the decision outcomes should be regarded as "the will of the majority" and not the "will of the people."[33]

Thus, Schumpeter, to his satisfaction, smashes the edifice of the classical doctrine into pieces. Given that Schumpeter was unable to pinpoint to whom exactly the "classical doctrine" was to be attributed, as mentioned earlier, and also given the simplistic and absolutist nature of some of the propositions he ascribed to this doctrine, in particular propositions A-E, there can be no doubt that his conception of the classical doctrine is but a straw theory that he himself set up in order to demolish. Without a doubt, his purpose in doing so was to show that the notions of governing based on the general will or common good of the people is untenable and thus must be abandoned in favor of his own theory that does away with it in order to define democracy as a value-free and logical method. Although a detailed examination of propositions F-O will not be presented here, some relevant discussions and criticisms will be offered below.[34]

The most striking feature of Schumpeter's rejection of the "classical doctrine" is the way in which it treats—and eventually dismisses as "myths"—the notions of the "common good" and the "will of the people." This is a crucial point, for his way of treating these notions led him to reject the idea of the rule by the people altogether. Briefly stated, in his polemic against the "classical doctrine," Schumpeter omitted from consideration the most essential component of democracy in the doctrine, namely the idea of the citizens' *direct* participation in decision-making. Recalling from Chapter 3, the fundamental idea of the classical doctrine (i.e., Rousseau's) was not that the people must be ruled in accordance with their will (the common will or the general will), but that the people *themselves* must do the ruling (i.e., the decision-making.)[35] What Schumpeter distorts in his polemic is the crucial point that the proof of the sovereignty of the people (or democracy) in the original idea (or the classical doctrine) is not that the will of the people is sovereign in the state (for this is only a consequence of their sovereignty), but that they *themselves* are the sovereign; that they themselves legislate the laws—or ratify "in person" the laws that have been drafted by their appointees on their behalf. Once in the position of governing or sovereignty, the people would have no will to govern in accordance with, except their own will, i.e., the most pronounced will that would emerge in the process of legislating the laws. Whether the will in question would be "natural" (i.e., preconceived and already-existing) or "composed" (i.e., a decision arrived at via deliberation in the process of legislating) is irrelevant. What matters is that the people *themselves* would have arrived at it.[36]

Moreover, one should take issue with the attributes of disinterestedness, moral and intellectual incompetence, moral corruption, and the absence of independent or definite wills that he ascribes to individual citizens. Despite Schumpeter's portrayal of these attributes as "true"—because he implies that they are verifiable by empirical observations—these attributes are more prejudicial and elitist in nature than scientific or objective as he intends to present them.[37] These negative attributes that are commonly ascribed to ordinary people will be examined in some detail in the following chapter.[38] For the time being it suffices to suggest that Schumpeter's rejection of the idea of the citizens' direct participation in decision-making (based on the empirically observed characteristics of the people in propositions J-M) is flawed and suffers from the all too familiar is-ought fallacy. Finally, it should be mentioned that Schumpeter's usage of the "general will" confuses the two senses of the word "general." One is not always sure what exactly Schumpeter means by a "general will." A will can be said to be general in that it is popular among the people, or in that it is comprehensive in what it covers or qualifies as the will of the people. Although it seems that Schumpeter is using the term in the first sense, the interpretation of the term "general" in its second sense is also plausible in some of his usage.

*   *   *   *   *

Now, Schumpeter's own conception of democracy. His "theory of competitive leadership," as he himself calls it, starts first with the fundamental proposition that democracy is to be treated merely as a *method* or a technique for decision-making, and *not an end in itself.*

> Democracy is a political *method*, that is to say, a certain type of institutional arrangement for arriving at political—legislative and administrative—decisions and hence incapable of being an end in itself, irrespective of what decisions it will produce under given historical conditions. And this must be the starting point of any attempt at defining it.[39]

And then vis-à-vis the classical doctrine that takes the people's decision-making powers and capabilities as the "primary" feature of democracy, and the role of the representation as "secondary," Schumpeter argues that:

> Suppose we reverse the roles of these two elements and make the deciding of issues by the electorate secondary to the election of the men who are to do the deciding. To put it differently, we now take the view that the role of the people is [or rather ought to be] to produce a government, or else an intermediate body which in turn will produce a government [the primary role of the people]. And we define: the democratic method is that institutional arrangement for arriving at political decisions in which individuals [the future representatives] acquire the power to decide by means of a competitive struggle for the people's vote.[40]

Thus, the question of democracy is the question of electing a representative government in a competitive process. Schumpeter claims that this definition "greatly improves the theory of the democratic process." Among the reasons he offers are the claims that his theory leaves an appropriate amount of room for the concept of leadership, which he believes plays an important role in democracy, and that his theory is compatible with individual freedoms.[41]

By all indications, Schumpeter's conception is built around a thick notion of leadership. The elected representatives or officials of the people are their leaders more so than being their representatives. Moreover, the competition among the candidates for the votes of the people is, in fact, the competition for their leadership. Schumpeter admits that this competition is vulnerable to the same sort of difficulties that dog the competition in the free-markets.[42] In order to assure that competition produces competent leaders, Schumpeter sees the need for a reliable stock of the "human material of politics."[43] There is a need for a special "social stratum"—or indeed a *"political or ruling class,"* as Carole Pateman puts it— that would rear "sufficiently good quality" individuals who would be *distinct* from the general population.[44] In reading Schumpeter on this matter one wonders whether Schumpeter is going beyond a simple notion of "natural aristocracy" or the principle of the distinction discussed earlier and whether he is hinting at establishing an institutional aristocracy. Among the abilities the leaders would possess, Schumpeter mentions the abilities of the "handling of men" and being "good tacticians" in addition to the qualities of character and intellect.[45] For Schumpeter, the fact that a candidate can survive and win the free competition for the political office is a good indicator that he has these abilities and qualities.[46]

In their position as rulers, the elected leaders would be assisted by technical experts and "the specialists' advice."[47] Moreover, the business of ruling and leadership will be further assisted by an efficient and powerful bureaucracy that has a "strong sense of duty" and can carry through the decisions made by the government. A strong bureaucracy is indispensable to Schumpeter's conception of democracy. He goes so far as granting the bureaucrats the power to "instruct the politicians who head the ministries."[48] Closely connected to the notions of strong leadership and bureaucracy is the idea that the range of the issues that would be decided democratically "should not be extended too far."[49] Not every decision must be subject to democratic procedure, which was a requirement in the classical doctrine.

As to the role to be played by the people in the division of political labor into electing the rulers and the business of ruling, in addition to electing the rulers and then leaving them alone to do the ruling, the people are expected to conform to the democratic culture by showing tolerance and respect for the opinions of their fellow citizens.[50] Although Schumpeter's theory refers to the task of producing a government as the "primary function" of the people, it turns out that this is their *only* function.[51] Beyond electing the members of the parliament or government officials (and rejecting them if they prove to be poor leaders), Schumpeter's theory assigns no other role to them. The people "do not decide issues," nor do they put forth the initiatives: "[i]n all normal cases the initiatives lie with the candidates who make a bid for the office."[52] Moreover, once the candidates are elected, Schumpeter wants the electorate to stay out of their way completely: "once they have elected an individual, political action is his business and not theirs. This means that they must refrain from instructing him about what he is to do."[53]

The elitist character and the aristocratic nature of Schumpeter's theory come through clearly. Decision-making is a prerogative of "high quality" individuals and expert bureaucrats, and not of the people. In his classification of the models of democracy, Macpherson has characterized Schumpeter's model as the "pluralist elitist equilibrium model."[54] In Macpherson's taxonomy, Schumpeter's model is the latest of three models of democracy the liberal society has developed since the late nineteenth century. Moreover, as Macpherson contends, Schumpeter's model or conception of democracy is nothing other than the adaptation of the market mechanism of competition to politics. And in this respect, the equilibrium model seems aptly to complement liberal-democratic capitalism: in the "political market" of "political goods," politicians (the political elite) compete (as individuals or as parties) for the votes of "political consumers" (the citizens) so that they can be elected as their representatives in the government and "do the deciding" for them.[55] According to Macpherson, exponents of this model of democracy claim that it "produces optimum equilibrium and some measure of citizen consumers' sovereignty."[56] Moreover, despite its pretence to being value-free—and rejecting the classical doctrine for being value-based—Schumpeter's own conception is highly value-laden. If the classical doctrine had fixated on making the will of the people sovereign in the state as its end, Schumpeter's theory aims at achieving a different set of goods, viz., stability and efficiency, both, especially the latter, being highly esteemed virtues of the capitalist society. That Schumpeter's theory of democracy is intrinsically bound to the value system of the capitalist order has been pointed out by many authors. In particular, Macpherson has likened Schumpeter's notion of the competition of the candidates for office to the competition of producers for market shares:

"[d]emocracy is simply a market mechanism: the voters are the consumer; the politicians are the entrepreneurs."[57]

In Schumpeter's theory of democracy, the representative government amounts to a government *of* the people (in the sense that they have elected it) but not a government *for* the people, for the government's *raison d'être* is not to serve the people or their common good, for after all, there is no such thing as common good. Rather, the government is elected in order to administer the society efficiently, and to do so based on a strict division of the political labor: the people elect the government and then stand back and let the elected officials exercise their decision-making powers—this would assure stability. Nor is the Schumpeterian government a government *by* the people, for the people are barred from it. In addition to the common good, Schumpeter also throws away the idea of the participation of the people in governing and substitutes for them his thick notion of leadership. Every intention of the theory of "competitive leadership" is to keep the people as far away from governing as possible. One can argue that, in addition to his contention that the idea of the common good was a myth and that citizens were politically disinterested and incompetent, for Schumpeter, a strong leadership of political elites and natural aristocrats was also warranted in order to compensate for the fact that, in his view, capitalism had undermined the hereditary aristocracy and the intellectuals who were instrumental in its rise to power in the earlier centuries.[58]

At the heart of Schumpeter's theory of democracy lies a deep-seated and cynical belief that "the people never actually rule but they can always be made to do so by definition."[59] One implication of regarding democracy as a "matter of definition" would be the claim that the main problem with the classical doctrine was that its definition was too unrealistic; and its assumptions about the people and their general will did not conform to the empirical facts about them, as well as to the problems involved in forming their general will. Moreover, Schumpeter was of the view that the people should not rule, not even indirectly. As far as one can see, there is no linkage between the decisions made by the elected officials, on the one hand, and the views and "wills" of the people, on the other hand. The strict division of political labor into electing (to be done by the people) and ruling (to be done by the elected officials and representatives) does not allow for the policies preferred by the people to have a role in decision-making. In short, there is no role assigned to the people once the election is over, direct or indirect.

*   *   *   *   *

With Schumpeter's "theory of competitive leadership," the *"liberal-democratic conception of democracy"* was born. Schumpeter's theory gained wide acceptance among the political scientists in the middle to late decades of the twentieth century and served as a fertile ground for the flourishing of numerous influential works that took his formulation as their point of departure. These works and their authors, in the words of Macpherson, "have amplified and supported it [Schumpeter's theory of democracy] by a substantial amount of empirical investigation of how voters in Western democracies actually behave and how existing Western political systems actually respond to their behavior."[60] These works, along with Schumpeter's own, contributed greatly to the widespread acceptance of the "liberal-democratic conception of democracy." With an eye on the role assigned to the individual citizens' participation, Pateman has examined four major works that contributed to the "liberal-democratic conception of democracy" during the two decades that followed the publication of Schumpeter's work (those of B. R. Berelson, Robert Dahl, G. Sartori, and H. Eckstein), and summarized what she called the "contemporary theory of democracy" as follows.[61]

> In the theory, "democracy" refers to a political method or set of institutional arrangements at national level. The characteristically democratic element in the method is the competition of leaders (elites) for the votes of the people at periodic, free elections. Elections are crucial to the democratic method for it is primarily through elections that the majority can exercise control over their leaders. Responsiveness of leaders to non-elite demands, or "control" over leaders, is ensured primarily through the sanction of loss of office at elections; the decision of leaders can also be influenced by active groups bringing pressure to bear during inter-election periods. "Political equality" in the theory refers to universal suffrage and to the existence of equality of opportunity of access to channels of influence over leaders. Finally, "participation," so far as the majority is concerned, is participation in the choice of decision makers. Therefore, the function of participation in the theory is solely a protective one; the protection of the individual from arbitrary decisions by the elected leaders and the protection of his private interests. It is in its achievement of this aim that the justification for the democratic method lies.[62]

Pateman has commented that the "contemporary theory of democracy" puts higher emphasis on the stability of the political system than Schumpeter's original formulation.[63] This observation is in agreement with Macpherson's repeatedly-made point in *The Real World of Democracy* that the value the liberal-democratic society and state have discovered in democracy is not the question of sovereignty and equality, nor the rule of the majority, but the *stability* of the status quo.

However, stability can only be guaranteed if the representative form of government could be justified not only as efficient and practical, but also as a just and reasonable form of government. In view of Schumpeter's contentions that the ideas of the common good and governing in accordance with the general will are nothing but myths, the legitimacy of legislative decisions and public policies then could only rest on the consent of the governed. But, the governed would give their consent only if they perceived the process of electing the government as fair and free, and regarded the legislative procedures of government as fair (insofar as they do not privilege, at least formally, any particular groups) and logical (e.g., efficient, workable, cost-effective, "scientific"). Thus, the questions of the *stability* of the representative government and the *legitimacy of procedures* that govern its operations are intrinsically connected to one another in the contemporary approaches to the question of democracy. These questions and their connections constitute the general subject matter and the main problematic of the liberal-democratic conception of democracy. In recent years, these questions have been addressed within the framework of "pluralism" as this concept, starting with the second half of the twentieth century, has managed to carve out a permanent space for itself in the American "public political culture" (mainly in the media), and in the academia (in its "normative," "political," "radical," and "neo-pluralist" forms).[64] It is in this light that the procedural notion of democracy, as well as the really-existing democracies, have been characterized by Robert Dahl as "polyarchy."[65]

The 1980s and 1990s witnessed the rise of new approaches to the question of democracy that extended the Schumpeterian idea of modeling democracy after market economics further, this time into the level of methodology, and applied its mathematical tools to the question of democracy. These new approaches materialized in what is often referred to as the "public choice theory," "social choice theory," and "rational choice theory." These theories have managed to cast a dark shadow on the idea of democracy as the rule by the people, in the same vein that Schumpeter's own theory did. In particular, the "theory of voting," a branch of social choice theory, has claimed that, even if we assume that individual citizens have definite and independent wills, there exists no large-scale decision-making schemes (such as voting) that can be used to channel the individual wills into a public or social will that can satisfy some criteria of fairness or logicity simultaneously.[66] (One should add that the late 1980s and 1990s also witnessed the rise of a trend in the liberal-democratic conception of democracy that ran in directions counter to some of the theories mentioned above. These theories will be discussed under the rubric of the theories of deliberative democracy in Part III.)

Notes

1. For a discussion of the contrasts between the liberal and democratic ideals see note 7 in Chapter 6.

2. John Gray provides a concise overview of the path that the revision of liberalism followed in England in the nineteenth century and the early twentieth century, and relates its development to the outbreak of World War I (Gray 1995, pp. 26-35).

3. One needs to add that in fact, these social legislations had begun in the 1910s with Woodrow Wilson's administration under the pressure, and fear, of a growing socialist movement in the United States. See also note 6 in Chapter 8.

4. Despite this negative meaning of the term in politics, classical liberalism made a come back in the 1980s as F. A. Hayek's *Constitution of Liberty* (1960), Robert Nozick's *Anarchy, State, and Utopia* (1974), and Milton Friedman's *Capitalism and Freedom* (1962) were read widely. The free-market conservatives on the right used the arguments

of these works against the welfare liberals (see Gray 1995, pp.36-41). Margaret Thatcher openly acknowledged her intellectual debt to Hayek's *Road to Serfdom.*

5. Recalling from Chapter 7, the latter principles were termed as the "principle of the distinction of representatives," the "principle of collective-procedural decision-making," the "principle of independent-virtual representation," and the "principle of frequent and competitive elections."

6. Schumpeter's resume included holding the position of finance minister in Austria in 1919.

7. Schumpeter (1942), p.244.

8. Ibid.

9. Ibid., p.295.

10. Ibid., p.284.

11. Schumpeter believed that the "ideology of the Rule by the People," or those of "the Will of the People" or "the Sovereign Power of the People" rose in the seventeenth and eighteenth centuries (Schumpeter 1942, p.247, also p.265).

12. Ibid., p.296.

13. Ibid., p.267.

14. Ibid., p.250.

15. Schumpeter mentions J. S. Mill only once (ibid., p.248) and Bentham twice (ibid., p.249, and p.260n).

16. Ibid., p.250.

17. Ibid., p.269.

18. Ibid., p.250.

19. Ibid., p.251.

20. Ibid., p.251.

21. Rousseau (1762), p.74. However, as was discussed in note 17 in Chapter 3, there seems to be an exception to this rule. In an obscure work titled *Considerations on the Government of Poland* completed in 1772 Rousseau took exception and accepted a delegatory form of government for Poland. Now, returning to Schumpeter's model of "classical doctrine," Rousseau's model of delegation in the exceptional case of Poland does fit the procedural or structural model Schumpeter attributes to the classical doctrine. Having granted this, one should still argue that Schumpeter should be faulted for misrepresenting or misunderstanding Rousseau's model of governing. Either it is the case that Schumpeter misrepresented the exception as the rule, or it is the case that Schumpeter was not familiar with this obscure work of Rousseau and he just assumed that this was Rousseau's view. Schumpeter does not give any indication which works of Rousseau he is using. He mentions Rousseau by name only once on p.249.

22. As one can infer from short descriptions of the Benthamite and (James) Millian models of democracy provided in note 46 in Chapter 8, these models were far from resembling any form of delegatory democracy. Moreover, their model was more concerned with keeping the government in the hands of the rich than with extending democracy to every member of the society. For details, see Macpherson (1977), pp.24-43.

23. Schumpeter (1942), all of the quotations have been taken from p.250; all of the italics are added in order to show that Schumpeter oversimplifies, and thus distorts, the "classical doctrine" to the point that it becomes untenable.

24. Ibid., p.251.

25. Ibid., p.251-52, italics added.

26. Ibid., p.252, original italics.

27. Schumpeter discusses this in pp.253-54 and pp.256-57. The individual citizens' wills are not independent because the citizens are susceptible to "pressure groups and propaganda" (ibid., p.254).

28. Schumpeter discusses this in pp.253-54.

29. According to Schumpeter, the corruption of the people is most evident when their "immediate and personal pecuniary profit[s]" are involved: "Experience that goes back to antiquity shows that by and large voters react promptly and rationally to any such chance. But the classical doctrine of democracy evidently stands to gain little from displays of rationality of this kind. Voters thereby prove themselves bad and indeed corrupt judges of such issues" (Schumpeter 1942, p.260. Also, see p.253).

30. "And so it is with most of the decisions of daily life that lie within the little field which the individual citizen's mind encompasses with a full sense of its reality. Roughly, it consists of the things that directly concern himself, his family, his business dealings, his hobbies, his friends, and enemies, his township or ward, his class, church, trade union or any other social groups of which he is an active member" (Schumpeter 1942, p.258).

31. Schumpeter believed that the ordinary citizens would fall for "groups with an ax to grind" or for "idealists of one kind or another or . . . [for] . . . people simply interested in staging and managing political shows" (Schumpeter 1942, p.263).

32. Ibid., pp. 254-55.

33. Ibid., p.272.

34. Some other relevant discussions and criticisms of these propositions are offered in numerous places in *Direct-Deliberative e-Democracy*.

35. As was discussed in Chapter 3, this view represents the positive interpretation of Rousseau's conception.

36. See Schumpeter (1942), pp.250-96. The characterization of "natural" is Schumpeter's own.

37. By claiming that these attributes are empirically verifiable and have withstood the test of history (e.g., p.260), Schumpeter implies that they have the status of scientific truths: these are the attributes that human beings have exhibited throughout history; therefore, this is how the people are.

38. See Chapter 4 of *Direct-Deliberative e-Democracy* for a more detailed examination of these attributes.

39. Schumpeter (1942), p.242, original italics.

40. Ibid., p.269.

41. Ibid., p.271.

42. Ibid., p.271.

43. Ibid., p.290.

44. Ibid., p. 290 and pp.290-91, respectively. Also, Pateman (1970), p.4 (note 1), italics added.

45. Ibid., p.288-89.

46. Ibid., p.289.

47. Ibid., p.291.

48. Ibid., p.293. As Schumpeter puts it: "It is not enough that the bureaucracy should be efficient. . . . It must also be strong enough to guide and, if need be, to instruct the politicians who head the ministries" (ibid.).

49. Ibid., p.291.

50. Ibid., p.294-95.

51. Ibid., p.273.

52. Ibid., p.282.

53. Ibid., p.295.

54. Macpherson (1977), p.77. The pluralist feature of the theory is entailed in that, given the diversity of the wants and needs, competing political parties offer "differently proportioned packages of political goods" for voters to choose (ibid., p.80). This is an equilibrium model because the government produced in this competition would be a stable one which "equilibrates demand and supply" (ibid.).

55. Ibid., pp.78-79.

56. Ibid., p.84.

57. Macpherson (1977), p.79. Macpherson intends this characterization as a general statement about the "pluralist elitist equilibrium model" of democracy and he takes Schumpeter to be its most important figure and systematizer (ibid., p.77).

58. This latter contention is advanced by Nicholas who bases it on a reading of Schumpeter's pp.417-18 (Nicholas 1981, p.116 and p.122).

59. Schumpeter (1942), p.247.

60. Macpherson (1977), pp.77-78.

61. Pateman (1970), pp.5-13. Pateman considers the following works in particular: B. R. Berelson's *Voting* (1954), R. Dahl's *A Preface to Democratic Theory* (1956), and *Hierarchy, Democracy and Bargaining in Politics and Economics* (1956), G. Sartori's *Democratic Theory* (1962), and H. Eckstein's *A Theory of Stable Democracy* (1966).

62. Ibid., p.14, emphases added.

63. Ibid., p.5.

64. Pluralism gained prominence in American political science in the 1950s and 1960s. The two building-block ideas of pluralism are "interest groups" and "power politics"—the idea that politics is about the competition among interests groups to acquire political power in order to further their self-oriented interests (which are posited as "subjective" in nature). A fundamental premise of the idea of "pluralist democracy" in its "classical" form is an equilibrium theory which assumes that competing interests in society have roughly equal opportunities to influence the public decision-making process, and their competitions are eventually resolved into a "self-correcting" balance of political power in the state. Cunningham has stressed the Madisonian origins and the Hobbesian connections of pluralism, as well as emphasizing the importance of the idea of leadership to the concept (which was also an important element in Schumpeter's theory) (Cunningham 2002, pp.78-80). See Held (1987), pp.186-201 and Cunningham (2002), pp.72-90 for short discussions of "classical" pluralism. Also see Cunningham (2002), pp.86-87 for a summary of four main criticisms often directed against pluralism. Despite the claims of earlier (classical) pluralism to being "realistic," empirically based, or descriptive, the premise in question represented only an ideal vision rather than an actual situation. Having granted this, the last two generations of pluralist theories have become "normative" or "prescriptive" as they have taken this idealized vision as an aim to be achieved. What Eisenberg (following McClure) calls "second generation" pluralists (and Held refers to as "neo-pluralists," e.g., later Robert Dahl and Charles Lindblom) are critical of the inequalities that work to the disadvantage of the less well-off. The "third generation" pluralists (again using McClure's classification) oppose restrictions on identity in politics or the projection and identification of fixed groups with fixed interests, and argue that these categories are contingently constructed. A good example here is Chantal Mouffe (regarded by Cunningham as a "radical pluralist"). See Held (1987), pp.196-220, Cunningham (2002), pp.184-97, Eisenberg (1995), pp.16-26.

65. According to Dahl, taken together, the following political institutions constitute a "polyarchy" (a system of rule by the many): "Elected representatives, . . . Free, fair, and

frequent elections, . . . Freedom of expression, . . . Alternative information, . . . Associational autonomy, . . . Inclusive citizenship" (Dahl 1998, p.92, figure 7). Dahl believes that these institutions are necessary to satisfy the "democratic criteria" of "effective participation," "voting equality," "enlightened understanding [of the issues]," "control of the agenda," and "the full inclusion" (ibid.). A more detailed and complete definition of polyarchy is to be found in Dahl (1956), pp.84-89. The practice of using the term "polyarchy" to describe the really-existing democracies began by Dahl and Charles Lindblom's work of 1953, entitled *Politics, Economics, and Welfare* (ibid., p.92).

66. A discussion of these theories, as well as a rebuttal to this claim of the theory of voting, is presented in Chapter 2 of *Direct-Deliberative e-Democracy*.

Chapter 10

Justifying Representative Government

As has been argued thus far, the "liberal-democratic conception of democracy" can be conceptualized as resting on six main principles: the new principle of universal suffrage, and the five principles of representative government that were imported from the liberal state. Beyond examining the arguments in support of the principles of representative government, and beyond considering in passing the claim that the representative government derives its legitimacy from being freely elected by the citizens, Part II has yet to offer an account of how the modern idea of representative government was *justified* in the beginning, and how it continues to be justified at the present time. This chapter will be devoted to this task. What follows is a brief examination of some of the arguments and fundamental assumptions that have often been used as justifications for the present form of representative government. As will be seen, a good number of these assumptions postulate ordinary citizens as ill-suited for performing the tasks of governing, thus implying that the representative form of government is a necessity. In what follows, these justificatory assumptions will first be categorized into seven distinct propositions and then will be discussed in some detail.[1]

Before categorizing these presuppositions, it is worth recapping the three features of the idea of representative government (in both the liberal and liberal-democratic states) that this part of the volume has thus far drawn out. First, in both states, the representative form of government functioned as an elitist apparatus. Second, since its inception, this form of representative government has been closely tied to the interests of the wealthy classes, and hence has provided them with easy access to the state. Finally, this form of representative government has been a conscious design intended to *disempower* citizens by "legitimately" transferring their legislative and executive powers to political elites who earn their elite status, and thus, acquire the right and authority to rule the people by winning competitive elections. It is a truism that the legitimacy of this form of representative government and thus the legitimacy of the rule by the political elite has always been justified in the name of the people. The assumption has always been that representatives function as the "trustees" of the people or that they rule by their consent and in "their interests."

Now, from the beginning in the liberal society, the arguments in favor of representative government have always been linked directly with arguments that put the case against the idea of the direct participation of everyday citizens in governing. A good example here is the way Madison approached the question of representative government, or the "republic," as he called it. Madison proposed a "republic" in which the function of governing was delegated to a small number of citizens, to a special "chosen body," as a substitute for, and in contrast to, the ancient idea of *direct* democracy, or "pure democracy" as he put it.[2] Despite this direct link between the arguments for the modern representative government, on the one hand, and those against the idea of direct democracy, on the other hand, one should bear in mind that the latter group of arguments are not specific to liberal and liberal-democratic societies. Rather, dating back to ancient Athens, these arguments have existed throughout the history of Western civilization. What is new is that, with the advent of the liberal society and the liberal state, the old arguments against the idea of direct participatory democracy have been resurrected and this time dressed up as arguments in support of the modern idea of representative government. Broadly speaking, the old arguments started from the presuppositions that ordinary people were unfit to govern. Proponents of the idea of the modern representative government, in addition to reviving these old presuppositions and contentions, have produced new arguments against the idea of direct democracy. In light of the large size and ever-rising complexity of the modern nation-state, as one of these new arguments goes, a direct system of government by the people would be impractical and inefficient. Another argument attempts to emphasize that ordinary citizens in the modern world have no interest in participating directly in governing. These new arguments, along with the old ones, have been omnipresent for the last two centuries or so.

What is unsettling is that, to many individuals living in liberal-democracies (at least to the ones living in the United States and Canada) these arguments appear as persuasive and intuitively valid. The omnipresence of the view that the idea of direct participatory democracy is not practicable should be attributed, in part, one can argue, to the fact that the arguments supporting this view are produced and disseminated on a regular basis in the mass media of communications. Citizens of liberal democracy are kept in the perpetual state of being bombarded with images and viewpoints that constantly pound the message, be it directly or indirectly, that the business of governing, or the universe of politics in general, is a complex and perilous one; that this universe is not for everyone and it takes individuals of certain capabilities, skills, and dispositions (intelligence, charisma, public speaking abilities, ability to persuade and make political deals, and also craftiness) to be effective in this world and survive in it as well.[3] The ubiquity of views and arguments against the idea of direct participatory democracy should also be attributed to the fact that the counter-arguments do not receive attention or favorable exposure in the mass media. Given this state of affairs, the status of the representative form of government has been elevated to that of inevitability and absolute necessity in the popular consciousness of the liberal-democratic society. One direct consequence of this has been that the idea of direct participatory democracy has been thrown into complete oblivion.

In what follows, the presuppositions that are often used as the premises of the arguments for the representative form of government—and at the same time, against the idea of direct democracy—will be categorized into seven distinct presuppositions, and then will be discussed in some detail. These presuppositions will be referred to here as the *assumptions of the impracticality, inefficiency, and undesirability* of the idea of direct democracy, and the *assumption of uninterestedness of ordinary citizens* in doing so.

I. *The logistic impracticality and inefficiency of direct rule by the people*: the immense size and complexities of the modern nation-state make it impractical (and inefficient), if not impossible, to assemble the entire citizenry for policy- or decision-making purposes.[4]

II. *The political impracticality or inefficiency of direct rule by the people*: the facts of the heterogeneity of the citizen body and the diversity of perspectives, and the existence of a wide plurality of conflicting interests in the modern nation-state, make practicing the idea of direct democracy inefficient, if not impossible.

III. *The political-technical and intellectual incompetence of ordinary citizens*: ordinary citizens lack the political-technical knowledge and the intellectual-rational power needed for sound and consistent social-political decision-making.[5] This problem is compounded in the face of the ever-rising complexity of governing in the modern nation-state.

IV. *The moral incompetence of ordinary citizens*: ordinary citizens lack the moral competence needed for sound social-political decision-making.[6]

V. *The inability of ordinary citizens to discern their own true interests and the true interests of the whole community*: ordinary citizens lack the wisdom, fortitude, and other necessary attributes needed to raise them above the level of their immediate interests and concerns and to help them develop a true understanding of their common interests.[7]

VI. *The political uninterestedness of ordinary citizens*: ordinary citizens are politically apathetic and lethargic. They are not interested in participating in politics and prefer to leave the responsibilities of governing to professional politicians.

VII. *The susceptibility of ordinary citizens to fall for demagogues and the fear of instability:* a system of direct-participatory democracy would either be an unstable republic on the permanent verge of collapse, thanks to constantly being undermined by the ordinary citizens' whims and prejudices, or would eventually fall prey to demagoguery.[8]

The proponents of the modern idea of representative government have succeeded in elevating these presuppositions to the level of plausible and intuitively valid truths and empirically verifiable facts. It is only in light of the acceptability of these propositions that arguments on behalf of representative government appear compelling. These presuppositions are examined at some length in *Direct-Deliberative e-Democracy*.[9] Here, it suffices to discuss some of them in passing. One can begin with presupposition VI, the *assumption of uninterestedness of ordinary citizens.* Limiting the scope of the discussion only to the American scene, there is no denying that this presupposition is based on empirical observations. Schumpeter put forth this claim forcefully and based it on empirical evidence.[10] Presupposition VI is an important argument in justifying the representative form of government: If everyday citizens are not interested in partaking in politics—and given that coercing them to do so would not be a good idea for they might sabotage the state—then obviously a representative form of government, by default, would be the only option.

Presuppositions III-V are *undesirability assumptions;* they postulate the ordinary citizens' participation in politics as undesirable. These presuppositions, no doubt, are based on a base opinion about, and low appraisal of, the ordinary citizens' capacities and potentials. Presuppositions III and V are in essence pseudo-epistemological assumptions about everyday people, while presupposition IV is a pseudo-moral justification for keeping them out of politics. Despite the fact that these presuppositions are more metaphysical than factual, they strongly appeal to intellectuals and elitist casts of mind. This appeal perhaps should be attributed to the still-powerful influence of the negative theories of human nature and negative views on the ordinary citizens' capabilities that originated in Plato and was reaffirmed by Machiavelli, Hobbes, Madison, and others.[11] Moreover, these presuppositions, as well as the theories of human nature that bestow intellectual respect on them, have aristocratic roots and are class-based in nature, a quality that has overwhelmingly tainted much of the intellectual production in the realm of social-political thought throughout the history of Western civilization.[12] As can be recalled, the arguments for Burkean and Madisonian conceptions of representation were also based on these presuppositions.[13]

As to presupposition II, it only suffices to say at this point that this proposition is a theoretical-ideological assumption, and, by all indications, it seems that it feeds primarily on suspicions about the negative effects of the ever-growing pluralism and diversity on the political stability of the liberal-democratic society. Presupposition I, on the other hand, is a purely empirical and objective one. It does not misrepresent the facts and is free of ideological prejudices and theoretical biases, a claim that cannot be made about presuppositions II-VI. Having granted this, it should be added that presupposition I is premised on a certain historically conditioned and limited understanding of the practice of direct democracy that can be stated as follows. Decision-making in a direct democracy would take place in physical assemblies wherein the decision-makers, the citizens, assemble together and hold face-to-face discussions that eventually arrive at a decision, either via some form of voting scheme or via reaching a consensus. This is in fact the model of the Athenian direct democracy that was also practiced in some isolated pockets in medieval Europe.[14] It is a truism that the immense size of the modern nation-state makes the ancient Athenian model of practicing direct democracy impractical. Finally, presupposition VII is a conclusion which is drawn on the strengths of presuppositions III-VI, and is the most metaphysical of them all.

These presuppositions, and the arguments they raise against the idea of direct democracy, constitute the justificatory foundations of the concept of representative government. Before closing Part II, it is of utmost importance at this point to offer a caveat on the anti-democratic portrayal presented of the representative government in Part II. This portrayal should be interpreted *only* as a critique and rejection of the *form* this government assumed in the liberal state and continues to assume *in the liberal-democratic* one, *and not* as a sweeping rejection of the *idea* of representative government. Any realistic theory of democracy committed to the original meaning of the idea must concede that some of the presuppositions listed above (especially I-II) pose serious challenges to the idea of direct-participatory democracy that cannot be met realistically, unless one is willing to accept some form of, or some role for, representative government. *Direct-Deliberative e-Democracy* introduces a special conception of representative government that could aid the goal of realizing the idea of direct-participatory democracy in the modern nation-state.[15]

Notes

1. A more detailed examination and critical analysis of these propositions are offered in Chapter 4 of *Direct-Deliberative e-Democracy*.

2. Madison in "The Federalist No. 10" (Hamilton 1961, p.133). As was argued earlier, by "a chosen body," Madison did not mean a body composed of a small number of citizens who were chosen as a representative sample (either by election or by lot), but a natural aristocracy whose members were superior to the rest of citizens, either due to their wealth or status, or due to their attributes of wisdom, virtue, or knowledge. The main reason Madison offered in support of his preference for a representative form of government (and an aristocratic one at that) was the impracticality of assembling all citizens in a large modern nation-state. Moreover, as to the distinction made by Madison between democracy and republic, Dahl comments that there was no precedence for making this distinction. His-

torical evidence and the usage of these terms at the time made these terms identical. In no prior historical time was the term republic understood to mean "a scheme of representation." Thus, Dahl concludes that Madison "made . . . [this distinction] . . . to discredit critics who contended that the proposed constitution was not sufficiently 'democratic'" (Dahl 1998,pp.16-17). In defense of Madison's identification of the "republic" with "a scheme of representation," one can evoke Quentin Skinner's argument to the effect that in the "neo-roman" republican movement that surfaced circa English civil war in the mid-seventeenth century, the participants in the movement (e.g., Nedham and Neville) favored the representative form of government over a system of the citizens' direct participation in governing (Skinner 1998, pp.31-32).

3. Interestingly enough, what is missing from this description are the attributes of moral fortitude, virtue, and honesty. Although these qualities are considered as virtues, it is generally accepted that they are not in high demand in the political universe.

4. As was mentioned earlier, this was the main thrust of J. S. Mill's argument against direct democracy (Mill 1958[1861], p.55).

5. McLean discusses a different version of this proposition which questions whether the ordinary people have the political wisdom to avoid making "inconsistent" and "unwieldy" decisions (McLean 1989, pp.109-111).

6. Here Schumpeter (1942) is a good case in point, e.g., p.257 and p. 260. According to Schumpeter, people can easily be corrupted when it comes to pecuniary matters.

7. Again, Schumpeter is a good example here. The ordinary people lack a "command of facts," a reliable "method of inference," and "the sense of reality [in them] is so completely lost" (Schumpeter 1942, ibid., p.261). As was seen above, J. S. Mill had a similar opinion of ordinary people.

8. For example, according to Schumpeter, people are "terribly easy to work up into a psychological crowd [even when they are not physically gathered] and into a state of frenzy in which attempt at rational argument only spurs the animal spirits" (Schumpeter 1942, p.257).

9. See the section titled "Justifying Direct Democracy" in Chapter 4 of *Direct-Deliberative e-Democracy*.

10. "And so it is with most of the decisions of daily life that lie within the little field which the individual citizen's mind encompasses with a full sense of its reality. Roughly, it consists of the things that directly concern himself, his family, his business dealings, his hobbies, his friends, and enemies, his township or ward, his class, church, trade union or any other social groups of which he is an active member" (Schumpeter 1942, p.258).

11. As was mentioned in Chapter 2, Plato had a low opinion of "common men." So was the case with Aristotle who believed "[m]en, when perfected, is the best of animals, but when separated from law and justice, he is the worst of all; . . . if he has not excellence, he is the most unholy and the most savage of animals, and the most full of lust and gluttony"—quoted by Loptson from *The Politics* I.2 (Loptson 1995, p.38). Machiavelli's view of the common men was worse. "Machiavelli conceived of 'the generality of men' as self-seeking, lazy, suspicious and incapable of doing anything good unless constrained by necessity" (Held 1987, p.44). (Held uses *The Discourses* as his reference. One should add that Machiavelli's characterizations of common men is much harsher in *The Prince*.) As to Hobbes, his portrayal of human nature in *Leviathan* presents Man as "profoundly self-interested, always seeking 'more intense delight' . . . [with] . . . 'a perpetuall and restlesse desire of power after power' . . . [sic]" (ibid., p.48). Finally, Madison worked with, as Held maintains, "Hobbesian assumptions about human nature" and regarded the

"search for preeminence, power, and profit" as "inescapable elements of the human condition" (ibid., pp.61-62).

12. Throughout this history, and again starting with Plato, "men," especially "common men," have been regarded as "beasts." It is worth noting here that Barber uses the phrase "politics as zookeeping" in order to characterize the conception of politics that takes Men as beasts ("sovereign lions, princely lions and foxes, bleating sheep and poor reptiles, ruthless pigs and ruling whales, sly polecats, clever coyotes, ornery wolves (often in sheep's clothing), and finally, in Alexander Hamilton's formidable image, all mankind itself but one great beast").—see Barber (1984), p. 20. (It should be added that Barber in this passage has in mind the "liberal democratic imagery.")

13. The same can be said about Schumpeter's arguments against the "classical doctrine" that relied on these presuppositions. See Schumpeter (1942), pp.253-54 and p.260 for some of his negative characterizations of individual citizens.

14. The Athenian version of direct democracy has been described by various authors, e.g., Dahl (1998), Budge (1996), and McLean (1989). Budge and McLean enumerate some of the shortcomings of this form of democracy.

15. See Chapters 3 and 4 of *Direct-Deliberative e-Democracy*.

Concluding Remarks to Part II

Part II provided an overview of the theoretical and historical developments that accompanied, and contributed to, the rise and consolidation of the liberal and liberal-democratic states. The common components of these two states, and the links that connected the latter to the former, were the five principles of representative government. The main difference between these two states, as well as the advantage of the latter over the former, was the principle of universal suffrage. The process of transition from the liberal state to the liberal-democratic one began gradually in the mid-to-late nineteenth century. World War I accelerated the process and by the time World War II came to an end, the transition from the liberal state to the liberal-democratic one was almost fully completed.

Reflecting at this point upon the numerous arguments, analyses, and historical data provided in Part II, two distinct features of the representative form of government stand out. The first is that in both the liberal and liberal-democratic states, the representative form of government was elitist by design, and thus was intended to keep ordinary citizens at a "safe" distance from the business of governing. The other distinct feature of this form of government was that it had strong affiliations with the interests of the "economic society," and thus with the interests of the wealthy classes who had privileged and dominant positions in it. Since its inception in the liberal state, the representative form of government has proven to be more attuned to the needs of the wealthy classes than to the interests of others, and accordingly has worked to the advantage of the former by permitting them to manipulate the political process for their strategic purposes. With the advent of liberal democracy and the welfare state, these affiliations did not disappear, nor did they weaken. What took place instead was that in light of the attention the state paid to the general welfare of society, these affiliations were rendered less visible, and thus appeared tolerable to the general public. Furthermore, these affiliations became less noticeable as they grew more intricate and became invisible as they were buried in the complexities of the functions of the new state. Viewed from the standpoint of the original meaning of the idea of democracy, these two features of the representative government in the liberal and liberal-democratic states should also be regarded as their main *democratic shortcomings*.

Moreover, with the birth and consolidation of the liberal-democratic state and society, a new conception of democracy began to take shape that was articulated first by Schumpeter's bold "theory of competitive leadership." This new conception was referred to as the "liberal-democratic conception of democracy" and was characterized as the idea of "rule by a freely and popularly elected representative government." In this new conception, the idea of rule by the people or the idea of their political empowerment that had predominated in the conceptions of democracy in the pre-liberal and non-liberal societies is relinquished altogether. The notion that democracy is about popular sovereignty, substantive equalities, and the direct involvement of all citizens in governmental affairs is jettisoned from the "liberal-democratic conception of democracy" completely. Instead, what has come to constitute the content of democracy in this conception is the understanding that the question of democracy is the question of the "freedom of choice" and "right to choose" of the people in the competitive marketplace of politics—that democracy is about giving people the right and freedom to choose their government officials. Democracy in this conception can also be understood as a "free" method for electing the officials of the representative government.[1] Understood as such, the fundamental problem of democracy is no longer the question of making the people their own rulers or empowering them politically, but rather that of freely, and thus "legitimately," electing the political elites who would rule the people. Thus conceived, democracy is *an institutional arrangement for "legitimately" detaching and alienating from the people the power to rule,* and handing this power over to the elected officials who would exercise it in the name of the people. Indeed, this "democratic" empowerment of the officials of the representative government is tantamount to *political disempowerment of the people and the betrayal of the original ideal of democracy.*

As will be seen in Part III, this perversion of the original meaning of the idea of democracy in the liberal-democratic conception has not gone unheeded. A wide spectrum of authors and political theorists coming from various theoretical and political backgrounds have raised numerous issues with this new conception throughout the second half of the twentieth century.

Notes

1. "Free" method in the sense that the individual citizens are not obstructed from expressing their preferences in elections, and that powers that be do not impose their own wills on the people.

Part III

The Case of the Late Liberal Democracy

Chapter 11

The Late Liberal-Democratic State: Crisis of Representative Democracy

Part II presented an overview of the theoretical and historical developments that accompanied the rise and consolidation of the liberal-democratic state. It also examined the "liberal-democratic conception of democracy." In bringing this volume to a close, Part III presents a short overview of some of the difficulties that the liberal-democratic state and the liberal-democratic conception of democracy have been grappling with for the last four decades or so. It is argued here that these problems, by and large, stem from the two democratic shortcomings of liberal democracy that were identified in Part II. In addition to presenting this overview, Part III also presents a critical analysis of two alternative theories of democracy— (theories of participatory democracy and deliberative democracy)—that have been developed as responses to the liberal-democratic conception in the past three decades or so.

    *   *   *   *   **

The talk of the "crisis of liberal democracy" began early on in the twentieth century when liberal democracy was in its infancy.[1] Despite the difficulties liberal democracies faced in the beginning, they managed to survive the challenge of fascism, and emerged victorious at the end of the World War II. The talk of crisis quieted down for two decades, and then suddenly returned and hit the Western liberal democracies in a blast as these societies woke up to the noisy decade of "the sixties."

In the decade that is often dubbed as "the sixties," the liberal-democratic societies of North-America and Western Europe were plunged into what could be characterized as a cultural-moral-political crisis. This crisis, in one respect, was a crisis of values, and in another respect, a "crisis of legitimation." The 1960s witnessed the breakdown of the "traditional" moral and political values, as well as witnessing strong expressions of distrust by the people (especially the younger generation) of their political systems and leaders. The sixties was the decade of sexual revolution, drugs, counterculture, student unrest, the anti-war movement, and the civil rights movement in the United States. This decade also witnessed the rise of alternative trends in the arts and culture, as well as a surge in popularity of alternative ways of thinking about, and approaching, social and political questions—most of which continued into the 1970s and well beyond. It is worth noting that, to the surprise of many, especially the "traditional" Marxists, the radicalism and social upheavals of the sixties erupted at the peak of the economic boom and the general prosperity of the post-World War II era that lasted until the early 1970s.

In some respects, the 1960s should be regarded as the decade of the "explosion of democracy." In the United States, people (mainly the younger generation and blacks) took to the streets and protested the war in Vietnam, and racial injustice at home. They condemned imperialist aggression abroad and sympathized with the poor and the oppressed at home and around the world. Protesters also exposed "the hypocrisy of the American ideals" and charged that "its democratic system [was] apathetic and manipulated rather than [being] 'of, by, and for the people.'"[2] Moreover, they challenged authoritarianism in government, the university, and in other spheres of life ranging from the workplace to the family, as well as questioning the legitimacy of the status quo as the government of the people. The drive to give the "democratic system" back to the people manifested itself in the form of popular movements that operated both inside and outside of the liberal-democratic establishment. On the outside, democracy was in the streets. It unleashed its force both in violent protests and in various acts of civil disobedience ranging from sit-ins to illegal quiet protests. Working within the establishment, the activists, on the one hand, attempted to democratize the internal structures and procedures of the establishment parties (e.g., demanding that the parties hold popular primaries for the party cadres and nominees), and on the other hand, organized mass voter registrations and mass membership drives (e.g., the mass voter registration of blacks in the southern states in the United. States) in order to break the stranglehold of the wealthy and the "industrial-military complex" over the parties.

For many who participated in the movements of the 1960s, politics was understood primarily as a way of life. Thus, as an integral part of their political work, the activists took it upon themselves to "expand and humanize politics itself."[3] This meant that they had to involve the everyday people in running the affairs of their communities. Thus, "the establishment of a democracy of participation" or "participatory democracy" and providing "the media for their common participation" was the order of the day in many activist circles.[4] It was taken for granted that, in order to participate, people had to be empowered—hence the slogan of "Power to the People!" was omnipresent in these circles. With democracy in the streets, while some progressive and radical activists were devoting their energies to community-building and developing alternative approaches to solving social problems at the community level, others were engaged in developing alternative ways of conceptualizing the idea of democracy and empowering the people in the realm of theory.

Under the influence of the social and political movements of the 1960s, and in conjunction with them, a new trend in alternative thinking began to take shape around the question of democracy. This trend as a whole was sharply critical of the theory and practice of liberal democracy. As will be seen in what follows, the trend survived the fall of the New Left and continued into the 1970s and 1980s, as well as into the 1990s.

One of the main criticisms of liberal democracy that surfaced in this new trend in democratic theory was the charge that liberal democracy had preempted the moral content of democracy, and had reduced it to a mere "method" or "procedure." It was argued that liberal democracy understood democracy primarily as a method for electing government officials, on the one hand, and as a procedure for reaching compromise among diverse and conflicting interests, on the other hand. Either way, democracy was understood instrumentally and narrowly. Consequently, it was argued that the moral content or substance of democracy had to be restored. Perhaps the resurgence of the idea in the 1960s, that democracy has a moral content and that the status quo ought to be judged by a set of moral criteria, should be traced to *The Port Huron Statement* of June 1962—the founding document of the Students for a Democratic Society (SDS). In many respects, this document can be regarded as the political and moral manifesto of most of the movements that mushroomed in the 1960s. *The Statement* can also be credited for laying down the principles for what soon came to be known as "participatory democracy":

> that decision-making of basic social consequences be carried on by public groupings; that politics be seen positively, as the art of collectively creating an acceptable pattern of social relations;
>
> that politics has the function of bringing people out of isolation and into community, thus being a necessary, though not sufficient, means of finding meaning in personal life;
>
> that the political order should serve to clarify problems in a way instrumental to their solution; it should provide outlets for the expression of personal grievance and aspiration.[5]

This is how the new alternative trend began.[6] Soon after, many authors began to pound the point that democracy has a moral content, which has been left out in the convenient marriage of democracy and liberalism.[7] In these works, the attacks on the moral deficiencies of liberal democracy were coupled with the presentation of a new conception of democracy that came to be known as "participatory democracy." The main contention here was that the moral content of democracy ought to be restored. On the strength of the works of Rousseau, G. D. H. Cole, and J. S. Mill, Carole Pateman argued that democracy was an end in itself. She argued that the participation of the citizenry, aside from being itself a moral principle, was a way of life that was conducive to educating citizenry, developing their human capacities, and realizing their potentials. As an alternative to liberal democracy, Pateman put forth a community- and workplace-based conception of direct participatory democracy. C. B. Macpherson, on the other hand, although agreeing with Pateman's contention that democracy was "a set of moral ends"—(as well as being "a quality pervading the whole life . . . a kind of *society*, a whole set of reciprocal relations between the people," and about developing the human individual)—mainly devoted his energy to retrieving the moral content of democracy in the realm of theory.[8] In the realm of practice, Macpherson envisioned a form of participatory democracy that was indirect in comparison to Pateman's.[9]

The criticism that liberal democracy is devoid of moral content and that it dilutes the citizens' sense of civic responsibility and social solidarity continued into the 1980s and 1990s. Communitarians in particular attempted to revive the "civic virtues" and criticized the primacy given to individual rights and freedoms over the considerations of the common good in liberal democracy. Michael Sandel's characterization of liberal democracy as "procedural republic" is one example.[10] For Sandel, this characterization is justified, for the "public philosophy" of the American liberal democracy "asserts the priority of fair procedure over particular ends."[11] As an alternative, Sandel put forth the republican political theory with its conception of freedom that "requires a formative politics, a politics that cultivates in citizens the qualities of character self-government requires."[12]

Another line of criticism that began to take shape in the 1960s, and continued well into the twenty-first century, was the argument that the liberal components of liberal democracy carry a heavier weight in the conception than its democratic counterparts. Alternately stated, it was argued that the democratic components in the conception lack real substance; that the liberal-democratic conception of equality is formal and not substantive. The dominant position of liberalism over democracy in the liberal-democratic theoretical conception is evident in the primacy the conception gives to choice and rights over equality and popular sovereignty. This is also evident in the individualistic and "anti-democratic bias" of the liberal-democratic notion of rights.[13] A good representative of this view in recent years is Benjamin Barber who contends "[l]iberal democracy is . . . a 'thin' theory of democracy, one whose democratic values are prudential and thus provisional, optional, and conditional—means to exclusively individualistic and private ends."[14] In liberal democracy, "democracy itself is never more than an artifact to be used, adjusted, adapted, or discarded as it suits or fails to suit the liberal ends for which it serves as means."[15] The alternative put forth by Barber is his theory of "Strong Democracy." Barber's Strong Democracy can be best described as a halfway compromise between direct (or "pure") democracy, on the one hand, and the liberal-democratic form of representative government, on the other hand. Strong Democracy is presented as the "government by all of the people some of the time over some public matters."[16]

Another author who has characterized the relationship between democracy and liberalism in the liberal-democratic conception as that of the subordination of the former to the latter is Andrew Levine. For Levine this "thoroughgoing subordination" is in part a consequence of the separation of the political and the state from the social, and the "devaluation" of the former spheres in the liberal-democratic conception.[17] This problem, along with a host of other problems, leads Levine to argue that liberalism and democracy are theoretically incompatible. According to Levine, "liberalism and democracy cannot . . . be combined satisfactorily."[18] And if they "cannot properly coexist, then one or the other must somehow give. It is, significantly, the democratic component that proves vulnerable and gives way in practice. Liberal democrats are, after all, liberals first, and democrats only reluctantly."[19] Among other problems that prevent the liberal-democratic conception from being a workable marriage between liberalism and democracy, Levine mentions the "conceptually problematic" nature of the notions of freedom, interests, and rational agency in the liberal-democratic conception.[20]

Apropos of the fundamental question of sovereignty, the liberal-democratic conception, as Levine argues, has broken completely with the tradition of democratic theory. The conception assigns a large degree of importance to the rights of citizens to elect their government. This emphasis on the right to elect works to the detriment of citizens' rights to take part in sovereignty in direct or substantive ways. Moreover, the liberal-democratic approach lacks the conceptual tools or lexicon needed to address the question of sovereignty, especially in the context posed by Rousseau. In light of the contention that the government rules by the *consent* of the people—or the contention that the people freely elect the government—the liberal-democratic conception brushes aside the question of sovereignty as irrelevant. Levine argues that in neglecting the question of sovereignty, liberal democracy avoids the question of the "nature of authority" and occupies itself only with that of the "proper limits" of authority.[21] The only interpretation of the notion of sovereignty that seems to have any relevance to the liberal-democratic conception is the idea of "consumers' sovereignty" which comes about as a result of imposing the economic models of market equilibrium on democracy.[22]

Moreover, given the strong association between capitalism and liberalism, on the one hand, and the dominant position of liberalism in the liberal-democratic conception, on the other hand, some critics have attributed the democratic shortcomings of the liberal-democratic conception to its strong attachment to property rights.[23] Marxists, especially those holding the "traditional" Marxist positions on liberal democracy, have pushed this argument to its furthest limits. As was noted in Chapter 5, Lenin and Trotsky rejected liberal democracy outright as a bourgeois form of democracy that facilitated the rule of the capitalist class (a minority) over the rest of the classes (the majority) by deceiving the latter to accept the legitimacy of the rule of the former. The central core of the Leninist-Trotskyite view reverberated in the works of some Marxist scholars in the period of the 1960s-1970s.[24] However, given that the Marxist critique of liberal democracy has always been tied to its critique of the capitalist state, in light of Nicos Poulantzas' analysis of the complexities of the capitalist state and Antonio Gramsci's contention that the power base of the capitalist class is diffused throughout the society and the state is only one pillar of this power, the harsh and often dismissive rhetoric of the traditional Marxist critique against liberal democracy has toned down considerably in recent years. This softening of the Marxist critiques has been accompanied by a rise of interest in analyzing the state as a complex entity, and in developing more elaborate theories for modeling the relations between the state and liberal democracy.[25]

*   *   *   *   *

Concisely stated, the attacks on the "liberal-democratic conception of democracy" that began to take shape in the 1960s—and continued into the present decade— charged that this conception, while short on moral content, is excessive in its attachments to rights in general, and property rights in particular; that liberal democracy is more in tune with the question of economic freedoms and rights (and thus with the question of private economic interests) than with questions of substantive equalities and public interests. Preoccupation with the former, as the argument goes, comes at the cost of undermining the importance of the latter. Alternately stated, the preoccupation with, and subservience to, the needs of the "economic society" is the main reason for liberal democracy's lack of adequate attention to the question of "civil society" and the issues of interest to the larger public—that range from the questions of substantive equalities to those dealing with "civic virtues."[26]

Putting aside the attacks on the "liberal-democratic conception of democracy" in the realm of theory, the clearest indications in recent decades that the democratic strain in liberal democracy is lacking substance, or that it is subordinated to the needs of the "economic society" (and thus to the interests of the capitalist class which has a privileged status in it), have been witnessed in the actual working of the liberal-democratic form of representative government.

During the last two decades of the twentieth century, as Marxist scholars began to shy away from directly linking liberal democracy to the interests of capitalist classes, numerous non-Marxist scholars and commentators stepped in to make similar claims and to advance the sort of criticisms that "traditional" Marxists used to voice. What was different about these claims and criticisms was that they were advanced on the evidence of growing empirical data that pointed to the presence of strong connections between the "organized money" or the "wealth," on the one hand, and "politics" and "democracy," on the other hand. With the century coming to a close, these criticisms grew sharper and louder. By many indications, these criticisms emerged as responses to the structural transformations that the major Western economies went through in the 1980s and the 1990s.

Briefly stated, as the post-World War II economic boom in the West slowly came to an end in the mid-to-late 1970s, the general sense of malaise that accompanied the resulting economic stagnation helped to set the stage for the rise of the right-wing and conservative politicians who successfully managed to lay the blame at the feet of the welfare-regulatory state. These politicians swept into political offices in the 1980s with their self-assumed "mandates" to bring down the welfare state, and to take the government out of the business of regulating the "economic society" (to "get the government off our backs," as Ronald Reagan once put it), and thus to give the control of the economy back to the markets. It was in the process of dismantling the welfare-regulatory state that the strong links between wealth and liberal democracy began to reappear. The "neoliberal" restructuring of the economy brought with it some flashbacks of the politics under the now-forgotten liberal state, where the wealthy and corporations had a free hand in ramming their self-interested agenda through the legislatures.[27] While the gradual dismantling of the welfare system went hand in hand with pushing the issues of interest to the public further down the list on the political agenda (including the eventual disappearance of concern for the poor from the agenda), with the deregulation of the economy, the concerns for the interests of the corporations and the wealthy topped the agenda. In the case of the United States, rollbacks toward the pre-New Deal era began full force under the Reagan administration and continued steadily under George Bush and Bill Clinton. Starting with George W. Bush's presidency in 2001, there is ample evidence that these rollbacks have begun to accelerate.

In recent years, criticisms of the neoliberal restructuring have crossed a new threshold. Alarmed with the anti-democratic implications of these rollbacks, a growing host of influential scholars, journalists, commentators, and public figures from all political spectrums in the United States are now voicing the serious concern that liberal-democracy is facing a serious threat. Some have warned of "The Corporate Takeover of Our Democracy," arguing that a "collusion between corporate interests and politicians" has torn down the "hard-won regulations that restrained the worst capitalist excesses."[28] Others have argued that the country is witnessing the reemergence of "plutocracy"; that it is suffering from "democracy deficit" and the "loss of national sovereignty": "America, in an ironical perversion of Lincoln's words at Gettysburg, . . . [has] . . . become a government of the corporations, by the corporations and for the corporations."[29] Still others have charged that the "[b]ig money and big business, corporations and commerce are again the undisputed overlords of politics and government" and that "[t]he unconstrained behavior of big business is subordinating our democracy to the control of a corporate plutocracy."[30] The traditional links between "big money" and "big business," on the one hand, and politics and government, on the other hand, have become so cozy and transparent in recent years that it is now commonplace for ordinary citizens to express the view that "the government is pretty much run by a few big interests looking out for themselves."[31] The popular discontent with recent developments has become increasingly widespread to the point that the three of the main campaigners in the 2000 U.S. presidential primaries felt compelled to address the question of "democracy gap" using strong terms and vivid expressions.[32]

In another ironical twist of Lincoln's words at Gettysburg, American democracy is amusingly characterized by one author as *"a government by the fund, of the fund and for the fund."*[33] The growing sentiments in recent decades that American democracy is a *Democracy Derailed*, a *Democracy for the Few*, a *Checkbook Democracy*, and the *Best Democracy Money Can Buy* fly in the face of the received view of American democracy as a "pluralist democracy" or a "polyarchy."[34] Now that the devolution movement to recreate the conditions of the pre-New Deal era is

pushing full steam ahead, it is becoming gradually evident that the latter characterizations of American liberal democracy are fathomable only if one takes the
existence of a welfare-regulatory state as a given.[35] In the absence of such a state,
as is becoming increasingly apparent, plutocracy, or perhaps "fund democracy"
would be more appropriate characterizations of a "democracy" in which contributing to the campaign funds of representatives, or lobbying their offices, results in
legislative decisions or executive actions that are favorable to the interests of those
who can and do afford to contribute or lobby.[36]

The other side of the coin of "fund democracy" is a phenomenon that can best
be represented by the phrase "audience democracy."[37]

Compared with the mid-twentieth century period when citizens believed that
they could exercise some degree of influence in formulating their parties' platforms or campaign themes (i.e., by volunteering in party activities such as political
campaigns or by taking a concern to a local party leader), citizens of the late liberal democracy appear and feel completely powerless and marginalized on matters
relating to influencing the political agenda.[38] In this respect, they have been reduced to mere audiences. As Manin puts it, "initiative of the terms of electoral
choice belongs to the politician and not to the electorate."[39] Manin rightly argues
that the role reserved for the electorate in "audience democracy" is a "reactive"
one.[40] In "audience democracy," the electorate does not get to "express" its own
views on what it considers to be the issues of public concern. The opinion polls
define the issues for the public; the mass media convince the public that opinion
polls reflect its sentiments accurately. Next, at the end of the chain, the representatives or candidates "tailor their images manipulatively in political advertising to
match public opinion poll results."[41] Finally, by means of complex techniques of
political marketeering and salesmanship, these images are presented to the electorate by their campaigns. Thus, as Manin puts it, "the electorate appears, above all,
as an *audience* which responds to the terms that have been presented on the political stage."[42]

           *   *   *   *   *

Now returning to the two main claims of Part II, "fund democracy," as the preceding overview strongly suggested, vividly exhibits the strong affiliations that Part II
claimed to exist between the "economic society," on the one hand, and the representative form of government in the liberal-democratic state, on the other hand.
After almost forty-five years (1935-1981), the shroud that conceals these connections has been ripped apart, thus rendering the connections visible again to the
naked eye of the general public. As to the other claim of Part II, that representative
government is elitist by design and is intended to keep citizens away from the
business of governing, one has sufficient grounds for making the argument that
this claim is equally applicable to the government of the late liberal democracy,
and perhaps more so in its present state of "audience democracy." What is mark-
edly different in this phase is that the "principle of distinction" of representatives
has become egregiously "commodified" and "commercialized." The commodification of representation takes the form of packaging the candidates and representa-

tives as political commodities.[43] It also takes the form of "manufacturing" images for them for the purpose of marketing them to the carefully-studied target groups.[44] The commercialization of representation, on the other hand, manifests itself in the form of advertising the candidates and representatives in the vast election markets run by a complex industry of pollsters, political consultants, "media specialists . . . image-makers . . . electronic wordsmiths, and . . . other manipulators of the public will."[45] In the political markets of "audience democracy," the distinction and the elite stature of this new breed of candidates and representatives stands out less by their grasp of issues, intelligence, knowledge, or their personal virtues, and more by their abilities and skills to raise funds, balance the competing demands of private-interest lobbyists, and manipulate the media and public opinion in projecting themselves as intelligent, trustworthy, and committed to serving the public, and thus as individuals of high and distinct stature.[46]

Moreover, with the increased role and influence of campaign funds and lobbying money in politics, the "principle of independent-virtual representation" has been weakened considerably, as the representatives' freedom of expression and action has become constrained by their ties to the corporate and other powerful private-interests. Another constraining factor on the independence of representatives has been the need to stay within the bounds of their party platforms, which largely stems from their need to have access to their parties' campaign machinery and funds. The conviction of the founding fathers of representative government, that the independence of representatives was a necessary condition for keeping them impartial, now appears as an ironic notion in "fund democracy." While representatives have managed to maintain their independence from the general electorate, the public, they have grown increasingly dependent upon the wealthy and corporations for their campaign funds.

Furthermore, under the ubiquitous pressure of funds and the influence-peddling of corporate and private-interest lobbyists, the "principle of collective-procedural decision-making" has become completely corrupted. The glorified claims of the architects of the idea of representative government, which held that representatives would engage in genuine deliberation on matters of interest to the whole nation in the legislative houses (Burke), or that they would strive to strike a balance among the competing (public) interests (Madison), now strike many citizens of the "fund democracy" as absurd and farcical. By all indications, what takes place in the legislative houses in the way of making laws cannot be properly regarded as genuine deliberation, nor can it be modeled as balancing the competing public interests. To the horror of many citizens of "audience democracy" who watch the spectacles from afar, the "debates" or "fights" that invigorate these houses can be more aptly described as an all-pervasive wheeling-dealing and logrolling that have the balancing of the lobbied interests as their objective.

Finally, in light of these considerations, the "principle of frequent and competitive elections"—(which in the beginning of the liberal state was regarded as a necessary and sufficient condition for *good* representation, especially by Madison)—has now been contorted into a principle that only serves to dignify the frequent and competitive fanfares of fund-raisings and public manipulations in "fund democracy." Nowadays, it is an established political "fact" that "any candidate with enough money and the right marketing strategy is guaranteed an audience to hear what he or she has to say and has a greatly increased chance of winning."[47] In major elections, as a general rule, the question of who wins the election often comes down to the questions of who spends the most campaign money (which is directly related to the question of whose campaign raises the most funds), and whose campaign is managed better (that is to say, the question of whose campaign does a better marketing job or a better job of "spinning" the facts, manipulating the public opinion and utilizing the media). Fanfares of frequent and competitive elections, and the heavy political marketing that accompanies them, not only confuse the average citizen and turn him into "a dazed man under siege," they also compound his apathy.[48] This, in turn, "nourishes the phenomenon of political marketing" that contributed to citizens' apathy in the first place.[49]

  *   *   *   *   *

The preceding sketches were intended as a brief overview of both the state of the "liberal-democratic conception of democracy" and the actual working of its representative form of government in recent decades. The sketches of the practices of the representative government in the United States in recent decades, and the characterizations of this government as "fund democracy" and "audience democracy," brought out the two main democratic shortcomings that the representative form of government has exhibited in both the liberal and the liberal-democratic states.[50]

  The rest of Part III will be devoted to a critical examination of two sets of theories of democracy that have emerged as responses to the problems of liberal democracy sketched above. The first is often referred to as theories of "participatory democracy"; the other is grouped under the rubric of theories of "deliberative democracy." Theories of participatory democracy were born in the midst of the movements of the 1960s and were keen on underscoring the moral content of democracy, as well as on emphasizing the importance of citizens' participation in the political life of society. In the following chapter, the views of the main proponents of these theories (in particular Pateman's and Macpherson's) will be discussed in some detail. While the 1960s and early 1970s witnessed the birth and surge of the theories of participatory democracy, the late 1970s and 1980s saw them wither away. Whether it was the hard economic times of the 1980s, or the final demise of what was left of the New Left, or the considerable weakening of the union movements in the United States, the idea of participatory democracy did not seem agreeable to the mood of the 1980s. Then came the 1990s with its alternative approach to the question of democracy, often dubbed as "deliberative democracy." The idea of deliberative democracy survived the 1990s and appears as alive and well at the opening year of the new century. A detailed examination of the theories of deliberative democracy will be offered in Chapter 13.

Notes

1. The first major declaration of this crisis was Carl Schmitt's *Crisis of Parliamentary Democracy* published in German in 1923 (translated to English in 1988). Approaching the problem from a fascist perspective, Schmitt charged that "the development of mass democracy has made argumentative public discussion an empty formality. Many norms of contemporary parliamentary law, above all provisions concerning the independence of representatives and the openness of sessions, function as a result like a superfluous decoration. . . . The parties . . . do not face each other today discussing opinions, but as social or economic power-groups calculating their mutual interests and opportunities for power, and they actually agree compromises and coalitions on this basis. The masses are won over through propaganda apparatus whose maximum effect relies on an appeal to immediate interests and passions" (Schmitt 1988, p.6). The main point of Schmitt's criticism was that politics was conflict-ridden and, given its essentially antagonistic nature (the friend-enemy conflicts), liberal-democracy would fail in its attempt to reduce the state and the function of politics to that of making laws through discussions among representatives in the parliament ("government by discussion"), and thus would fail to become a pluralist democracy. In Schmitt's estimation, liberal democracy had already turned into an instrument of the interests of the powerful social and economic groups who had forged a political unity among themselves against the weaker social and economic strata. See Mouffe (1993) and Hirst (1990), especially pp.105-37, for further discussions of Schmitt's view on liberal democracy. See also Cunningham (2002), pp.191-97 for a short overview of Mouffe's work.

2. These characterizations were used in *The Port Huron Statement,* the founding document of the Students for a Democratic Society (SDS) (included in James Miller (1987), p.330)—*The Statement* is included as an appendix in James Miller's *"Democracy Is In the Streets": From Port Huron to the Siege of Chicago* (1987), pp.329-77. As will be noted later in this part of the volume, *The Port Huron Statement* should be regarded as the founding moral and political document of the movements of the 1960s.

3. The phrase inside the quotation marks is borrowed from Seth Moglen in *Out of Apathy: Voices of the New Left Thirty Years On,* Archer, *et al.* (1989), p.6.

4. The phrases inside the quotation marks are borrowed from *The Port Huron Statement* (in James Miller 1987, p.333). Cunningham notes that the term "participatory democracy" was coined by Arnold Kaufman (one of the advisors to SDS) in 1960 (Cunningham 2002, p.122).

5. Ibid.

6. See Mansbridge (1980) for some case studies of the alternative approaches to practicing democracy that were influenced by the movements of the 1960s and 1970s. The work also contains some discussions about the influence of *The Statement* on these movements, e.g., p.21, pp.244-45, and pp.299-300.

7. The most important works here were Macpherson's *The Real World of Democracy* (1965) and Pateman's *Participation and Democratic Theory* (1970), and later Macpherson's *Democratic Theory: Essays in Retrieval* (1973) and *The Life and Times of Liberal Democracy* (1976).

8. Macpherson (1977), p.78, and pp.5-6, respectively, original italics.

9. The views of Pateman and Macpherson will be examined in some detail in the following chapter.

10. Sandel (1984), p.81 and Sandel (1996), p.4.

11. Sandel (1984), p.4. It should be added that here Sandel has Rawlsian liberalism in mind.

12. Sandel (1996), p.6.

13. Barber (1998), p.81. Barber uses this characterization in the American scene and within the context of discussing the debate between Federalists and Anti-Federalists.

14. Barber (1984), p.4.

15. Ibid., p.20.

16. Barber (1998), p.74. Full quotation: "not government by all the people all of the time over all public matters, but government by all of the people some of the time over some public matters." Barber believes that liberalism is indispensable to the project of democracy as well as being far more suitable as a foundation for the project than its rival Marxism, e.g., in the chapter titled "Why Democracy Must Be Liberal: An Epitaph for Marxism" in *Passion for Democracy* (1998). Barber's Strong Democracy will be discussed in further detail later in the following chapter.

17. Liberal democracy takes the "[c]ivil society, not the state . . . [as] . . . the sphere of human self-fulfillment, the area where human energies are best and most productively expended. To devote time and attention to legislation, then is to drain energy away from where it is best employed" (Levine 1981, p.144). Moreover "[I]f the state is only a necessary evil [as liberal democracy takes it to be], why expect it to further individuals' interests? It is perhaps a sense of this tension that accounts, at least in part, for the thoroughgoing subordination of democracy to liberalism that pervades so much liberal-democratic theory and practice" (ibid., pp.31-32).

18. Levine (1981), p.3, also p.203. Levine is a pessimist. He regards the "actual liberal democracy" as an "apparent 'solution' to an insoluble problem," viz., the problem of political authority (ibid).

19. Ibid., p.145.

20. Ibid., p.203.

21. Levine (1981), p.13.

22. Macpherson uses the phrase "consumers' sovereignty" as a characterization of how citizens' participation in politics is conceptualized in the Schumpeterian "pluralist elitist equilibrium model" of liberal democracy. In the "political market" of "political goods," entrepreneurs (the politicians) compete for the votes of "consumers" (the citizens) in order to be elected as their representatives in the government and do the deciding for them (Macpherson 1977, pp.78-79). It should be noted that in Macpherson's retrieval project, the notion of sovereignty did not play an important role. Macpherson was more interested in retrieving the egalitarian component of democracy; the same is true of Pateman's work.

23. A good example here is Jennifer Nedelsky's *Private Property and the Limits of American Constitutionalism* (1990). Nedelsky argues that the protection of private property has been the focus of securing the individual's rights and liberties in the American Constitution. Another relevant example here is Bowles and Gintis' *Capitalism and Democracy* (1986). According to the authors, liberal democracy is constantly badgered by the "clash" between the "property rights"—which enjoy a dominant position in the existing liberal democracies—and the "democratic personal rights" which attempt to challenge them.

24. Examples here are Ralph Miliband's *State in Capitalist Society* (1969), pp.238-39, and pp.242-43; Paul A. Baran and Paul M. Sweezy's *Monopoly Capital* (1966), pp.155-59; Ernest Mandel's *Marxist Theory of the State* (1971).

25. The complex Marxist models of the relations between the state and liberal democracy allow room for recognizing the contingencies and varieties of the state forms and operations. David Held provides a summary of the debate between Poulantzas and Mili-

band on the question of the nature and functioning of the capitalist state in the early 1970s, as well as a discussion of the reopening of the debate in the 1980s (Held 1987, pp.206-14). A recent discussion of the Marxist theories of the state can be found in Paul Wetherly's "Marxism, Liberalism and State Theory" (see Cowling 2000, pp.146-60).

26. Two notes are in order here. First, this summary of criticisms against liberal-democracy does not include the attacks launched by libertarians, conservatives, or free-market ideologues who believe the democratic component in the liberal-democratic amalgamation works to cripple its liberal component. Second, the terms "economic society" and "civil society" were defined tentatively in note 1 in Chapter 8, and will be discussed in detail in Chapter 4 of *Direct-Deliberative e-Democracy*.

27. This is how Noam Chomsky defines the term "neoliberal" or "Washington consensus" as he calls it:

> The basic rules, in brief, are: liberalize trade and finance, let markets set prices ("get prices right"), end inflation ("macroeconomic stability"), privatize. The government should "get out of the way"—hence the population too. . . . .

> The "principle architects" of the neoliberal "Washington consensus" are the masters of the private economy, mainly huge corporations that control much of the international economy and have the means to dominate policy formation as well as the structuring of thought and opinion (Chomsky 1999, p.20).

28. The phrase inside the first set of quotation marks is the subtitle of a chapter in Arianna Huffington's *Pigs at the Trough: How Corporate Greed and Political Corruption Are Undermining America* (2003). The other phrases inside the quotation marks are quoted from the same chapter, pp.77-78.

29. Phillips (2002), p.xi, p.xiii, pp.407-20, in particular p. 413 and p.417, and p.xvi. Phillips' concern with the loss of national sovereignty, as well as with the loss of the power of local governments in the U.S., is in part related to his critique of globalization. Phillips is borrowing the sentence in quotation marks from historian Arthur Schlesinger.

30. The first sentence inside the quotation marks is taken from Bill Moyers commentary on American National Television (PBS), cited in Phillips (2002), p.xvi. The second sentence is quoted from Ralph Nader (2002), p.61.

31. The phrase inside the quotation marks is borrowed from Craig (1996), p.46, who quoted it from a survey conducted in 1988. Among the recent books that have taken critical perspectives on the links between wealth and politics in the United States, one should mention the following: *Selling Out: How Big Corporate Money Buys Elections, Rams Through Legislation, and Betrays Our Democracy* (Green 2004), *Thieves in High Places: They Have Stolen Our Country and It's Time to Take it Back* (Hightower 2003), *Are Elections for Sale?* (Donnelly 2001), *The Corruption of American Politics* (Drew 2000, especially Chapter 4), *Speaking Freely: Former Members of Congress Talk about Money in Politics* (Schram 1995), and *Politics and Money* (Drew 1983). The following two books explore some aspects of the popular discontent with the American government in recent years: *What Is It about Government That Americans Dislike?* (Hibbing 2001) and *Broken Contract? Changing Relationships between Americans and Their Government* (Craig 1996).

32. The phrase "democracy gap" is borrowed from Ralph Nader (2002), p.56. The following three statements make the point vividly. First, Ralph Nader:

> The unconstrained behavior of big business is subordinating our democracy to the control of a corporate plutocracy that knows few self-imposing limits to the spread of its power to all sectors of our society. Moving on all fronts to advance narrow profit motives at the expense of civic values, large corporate lobbies and their law firms have produced a commanding, multifaceted, and powerful juggernaut. They flood public elections with cash, and they use their media conglomerates to exclude, divert, or propagandize. . . . . By their control in Congress, they keep the federal cops off the corporate crime, fraud, and abuse beats. They imperiously demand and get a wide array of privileges and immunities: the tax escapes, enormous corporate welfare subsidies, federal giveaways, and bailouts. . . . (Nader 2002, pp.327-28)—quoted from his speech announcing his candidacy for the Green Party's nomination for president.

The U.S. Senator John McCain of Arizona in the Republican primaries:

> We have squandered the public trust. . . . We defend the campaign finance system that is nothing less than an elaborate influence peddling scheme in which both parties conspire to stay in office by selling the country to the highest bidder (quoted in Nader 2002, p.61).

The former U.S. Senator Bill Bradley of New Jersey in 1996:

> Money not only determines who is elected, it [also] determines who runs for office. Ultimately, it determines what government accomplishes—or fails to accomplish. Congress, except in unusual moments, will listen to the 900,000 Americans who give $200 or more to either campaign ahead of the 259,600,000 who don't (quoted by Phillips 2002, p.405).

33. Mukherjee (2000), p.2, original emphasis.

34. The italicized phrases are titles of books by Broder (2000), Michael Parenti (1996), Darrell M. West (2000), and Greg Palast (2002), respectively. The main thrust of the arguments in these books is that "moneyed" and special interests manipulate and corrupt the elections and decision-making processes in American democracy, be it at the federal level, or the local/state levels. (Palast's focus is on the Florida debacle in the 2000 presidential elections in the United States.) These are among many books (mostly by commentators and journalists) that have appeared in recent years with the intention of taking the problem to the public. To the list above, one should also add Judis' *Paradox of American Democracy* (2000), which in part puts the blame on elites and the "enlightened political establishment" who offer their talents and expertise as services to special interests. Finally, one should mention Cambridge University economist Noreena Hertz's *Silent Takeover: Global Capitalism and the Death of Democracy* (2001). Hertz argues that the state's tilt toward serving the corporations works against its commitment to being the government of the people.

35. The concept of "pluralism" gained currency in American political thought in the 1950s and 1960s when a regulatory-welfare state was firmly established; and the defenders of the status quo then could argue that the existence of social safety nets meant that the interests of the poor were also represented, and also that regulations protected the interests of the working- and middle-classes against the wealthy. But even then, given the all-prevailing unequal dispersion of political resources or unequal access to the corridors of power, the pluralist idea of democracy was more of an idealized notion than empirically present—and this is despite the fact that pluralism presented itself, by and large, as an empirical or explanatory theory. The organized "interest groups" that competed among each other in influencing public decisions mainly represented the interests of a small minority of Americans (those belonging to the upper echelons of society). See Held (1987), pp.186-220 and Cunningham (2002), pp.72-90 for short discussions of "classical" pluralism. Also see Cunningham (2002), pp.86-87 for a summary of four main criticisms often directed against pluralism; Grady (1993), pp.32-52, for a critique of the liberal foundations of pluralism; and Ryden (1996), pp.67-85 for a general critique of the concept from the perspective of the difficulties it poses for the American constitutional law. (Ryden's work also contains a short overview of the responses offered by the "neopluralists" (e.g., Iris Marion Young, Lani Quinier, and Robert Grady) to the problem of "classical" pluralism, ibid., pp.88-96.)

36. Attempting to sound non-judgmental on what function lobbying serves in American democracy, Thomas R. Dye in his sixth edition of *Who's Running America?* refers to lobbyists as "fixers" in Washington "who offer to influence government policy for a price" (Dye 1996, p.133). According to Dye, the most important thing the lobbyists do is to "provide their clients with *access* to the corridors and cocktail parties of power" (ibid.). What Dye does not mention is that it is mostly the big and organized private interests who do the lobbying, and can afford the "price" of doing it. According to the information provided by "The Center for Responsive Politics," one can easily see that roughly 85 percent of the total amount spent for lobbying in 2000 (about $1.45 billion) was paid by big business and other private interests—the remaining 15 percent was spent by "ideological/single-issue," the "labor," and "other" groups. (The data provided for 1996-1999 show similar percentages.) Source: Internet site <wysiwyg://52/http://www.opensecrets.org/lobbyists> (31 May 2002).

37. The phrase "audience democracy" is borrowed from Manin (1997), p.218 and p.223.

38. One should be quick to add that this belief of citizens in the earlier part of the twentieth century was more of a perception than a reality. As Macpherson has argued, with the transformation of the class parties to mass parties, beginning with the dawn of the liberal-democratic era, the party system as a whole became "necessarily less responsible to the electorate. . . . Effective organization required centrally controlled party machines. . . . The main power fell to the party leadership" (Macpherson 1977, pp.67-68).

39. Manin (1997), p.223.

40. Ibid.

41. The phrase in quotation marks is borrowed from Yankelovich (1991), p.216. Candidates or representatives interpret the polls in ways that support their own policies, or exploit the sentiments expressed in polls to give popular "spins" to their own political programs.

42. Manin (1997), p.223, original italics. In recent years, one persistent feature of "audience democracy" has been the familiar phenomenon of voter "apathy" and "popular alienation from politics" (Phillips 2002, p.417). Many citizens, as Michael Parenti puts it

"doubt they have any ability to make government more responsive to their needs" (Parenti 1996, pp.195-96).

43. The notion of the commodification of representatives used above is developed after Jeffrey H. King's notion of "commodification of politics" in his "Public Policy, Commodification and Democratic Representation" (in Kochler 1987, p.103).

44. "At the core of candidate-manufactured images," according to Newman, "is the attempt to manipulate and control media coverage to paint the best possible television face for a candidate and, at the same time, to mold an image consistent with the appeal that the candidate wants to use to win over voters" (Newman 1999, p.106).

45. The phrase inside the quotation marks is borrowed from William D. Perdue in his "Crackpot Democracy: Reification and Ideology" (in Kochler 1987, p.103).

46. Ability to attract money, according to O'Shaughnessy, is the "primary task and central occupation" of a successful candidate (O'Shaughnessy 1990, p.9).

47. Newman (1999), p.3.

48. The phrase inside the quotation marks is borrowed from O'Shaughnessy (1990), p.243. As to the confusion of citizens, Newman argues that "[c]onstructing political meanings and realities out of the myriad messages coming out of a political campaign is becoming very difficult for the average citizen" (Newman 1999, p.106).

49. This argument is made by O'Shaughnessy (1990), p.243.

50. As can be recalled from Part II, the first of these shortcomings is the thesis that the representative form of government is a conscious design to keep citizens at a safe distance from the business of governing, and the other, the constantly recurring theme that the interests of the wealthy (and well-funded or well-organized interest groups) occupy a privileged place in the representative form of government. Moreover, given the critical nature of this study, it should not come as a surprise that the account presented above of the working of the liberal-democratic form of representative government runs counter to images of this government often portrayed by the "mainstream political scientists and other apologists" of the present system who try, in Michael Parenti's words, "to transform practically every deficiency in our political system into a strength" (Parenti 1996, p.vii). In his critique of the "apologists," Parenti continues: "They would have us believe that the millions who are nonvoters are content with present social conditions, that high-powered lobbyists are nothing to worry about because the president is democratically responsive to broad national interests" (ibid.).

Chapter 12

Theories of Participatory Democracy

David Held's discussion of "participatory democracy" in his comprehensive work *Models of Democracy* focuses on the works of Carole Pateman, C. B. Macpherson, and Nicos Poulantzas as the three prominent theorists of participatory democracy. The latter author belonged to the tradition of Western European Marxism and took issue with the state-based notion of socialism which dominated the theory and practice of Russian Marxism in Eastern Europe at the time. According to Held, Poulantzas emphasized the need for making the parties, state bureaucracies, and parliaments accountable and more open to the public. Poulantzas also emphasized the importance of developing forms of participation and struggle at local levels.[1] In the North-American scene, the idea of participatory democracy found its fullest expression in the works of C. B. Macpherson and Carole Pateman. The discussion of the idea of participatory democracy presented in this chapter is based primarily on the works of these two authors.

The notion of participation in the works of these authors was understood mainly as the participation of citizens in political decision-making. In order to realize this idea, it was taken for granted that there was a need for both a political process and an accompanying culture that would facilitate the convergence, and to some degree the aggregation, of the participants' opinions and preferences into some sort of mandate(s) that could be presented to the executive strata of the social, political, or economic organizations to which the participants belonged. Moreover, it was understood that there was also a need for some sort of mechanism and process through which the participants could hold the executive bodies accountable. This required, or implied, that members or participants had to have some sort of involvement or input into the workings of the executive branches of their organizations. The focus of Pateman's participatory democracy was on micro levels of the workplace, where the workers were to be involved in various capacities in decision-making as well as participating in the overall management of the factory or firm. Pateman found theoretical support for her theory of participation in the works of Rousseau, John Stuart Mill, and G. D. H. Cole. Building on G. D. H. Cole's views, Pateman developed a full theory of worker self-management. Moreover, following Rousseau, she emphasized the politically educative role of participation. Given that Pateman's model of participation was devised primarily for the micro levels, it had to remain, as Carol Gould has pointed out, "at the level of face-to-face relations among individuals and . . . [could] not extend to a consideration of social institutions [at broader national levels]."[2]

One needs to stress that Pateman's model was not fully participatory, and thus not an anti-representational one. A careful reading of Pateman's *Participation and Democratic Theory* suggests that, although she identified democracy with participation, she did not lose sight of the role, albeit limited, of representation. She favored a specific form of actual representation that she, following Cole, termed "functional representation."[3] This interpretation of representation in her model limits the power and role of the representatives to "some well-defined function[s]" of governing.[4] The structure of participation preferred by Pateman was Cole's model for industry that can be characterized as a pyramid. The pyramid took the ground of the workplace as its base. Pateman was not clear on the structure of power in the pyramid or on how high the pyramid rose. By all indications, however, it seems that in her model the participants at each level of the pyramid would directly participate in decision-making in the affairs of their level and elect the members of the next higher level. Moreover, one can further speculate that the multitudes of local/small pyramids would combine into one giant national pyramid, and the tip of the pyramid would constitute the ruling stratum of society. Finally, Pateman was not clear on how she perceived the relations between the pyramid power structure and the general political structure of society. Given what the reader was presented with, one can only speculate that she had a one-party system in mind which was similar to the system of the Soviets (councils) in the early years of the Russian Revolution. The major difference here seemed to be that in her theory the participants had much higher levels of political education and also that the participation took place in a truly free and democratic atmosphere. One should also note that Pateman's notion of participatory democracy, as her pyramid model indicates, did not draw a clear distinction between the economic sphere, on the one hand, and the political realm, on the other hand. This, however, should not be taken as an indication that politics was subjugated to economics in Pateman's conception. On the contrary, there is ample evidence to support the view that for Pateman, politics had a more fundamental role than economics; and as Jon Elster has noted, politics in her model of participation was taken as an end in itself.[5]

Macpherson's idea of participation, on the other hand, was presented in more general terms in comparison to Pateman's. And for this reason, it could be interpreted as having applicability in most social and economic organizations and institutions. Having stated this, one needs to add that Macpherson's writings on the question of models and institutional designs that could facilitate the citizens' participation in these organizations and institutions are scanty and vague. By all indications, it seems that by participation, as Carol Gould observes, Macpherson mainly meant "participation in political decisions about *economic* life."[6] That is to say, Macpherson's idea of participatory democracy was basically geared toward the advocacy of "increased *political-democratic control over the economy*."[7] Thus, compared to Pateman's attention to micro aspects of participatory democracy, Macpherson's idea of participation appears as deficient in elaborating how the workers and their associates could participate in democratic decision-making or how they could take part in managing and controlling the workplaces and firms.[8]

The view that Macpherson's theory of participatory democracy was mainly concerned with democratic control over the *economy* has also been expressed by Peter Lindsay. However, Lindsay is somewhat misleading in his affirmation of this point. Lindsay first gives the impression that there is strong evidence to suggest that Macpherson was clear on what constituted *political* participation, and that this type of participation was a strong component of his theory. But soon, he quotes Macpherson to the contrary. According to Lindsay, Macpherson was "a bit vague on the issue of what sorts of participation, *beyond the political*, are needed"—which implies that political participation was the main sense of participation in Macpherson's mind.[9] Lindsay takes Macpherson's allusion to the fact of the "rise of neighborhood and community movements and associations formed to exert pressure to preserve or enhance those [human] values against the operations of what may be called the urban commercial-political complex" in American cities, as a proof, and an example, of Macpherson's model of non-economic participation. However, he soon quotes Macpherson to the contrary, that "these movements, 'will not do the whole job . . . [for what is needed is the society's] democratic political control over the uses to which the *amassed capital* and the remaining natural resources of the society are put'."[10]

However, notwithstanding the fact that Macpherson's idea of participatory democracy is focused on the democratic control over the economy, and thus on "economic equality," one finds in a brief passage in his *Life and Times of Liberal Democracy* a scanty discussion of the application of participatory democracy to the political arena at macro levels. Speaking within the context of the question of electing the national government in a participatory scheme, there Macpherson proposed using a participatory pyramid model for reforming, and thus democratizing, the internal structures of the existing political parties.[11] One advantage of Macpherson's pyramidal model over Cole-Pateman's is that he combined the participatory pyramid model with the multi-party system.[12] His participatory model, as he suggested, would "keep the existing structure of government [intact], and rely on the parties themselves to operate by pyramidal participation."[13] It seems that this was the extent of Macpherson's attempts to consider the idea of participation in the sphere of the political at the macro level.[14] Moreover, by all indications, it seems that Macpherson's theory also relied on the existing representational model of North-American liberal democracy. As he put it, "[w]e cannot do without elected politicians. We must rely, though we need not rely exclusively, on *indirect* democracy. The problem is to make the elected politicians responsible."[15] One should note that Macpherson's strategy to "make . . . politicians responsible" seems to be consistent with his "praise," as Lindsay puts it, for the neighborhood and community movements and associations, which were assumed to function primarily to "exert pressure" on municipal structures rather than to "seek to replace" them.[16]

On the basis of this brief overview of Macpherson's and Pateman's theories of participatory democracy, one can argue that both theories fall short on addressing questions that deal with developing political processes and organizational models suitable for practicing the idea of participatory democracy in *macro* political arenas. Notwithstanding this shortcoming, theories of participatory democracy fared better in another area which was their true realm of interest. The most important component of the project of participatory democracy was the task of rescuing the true meaning of democracy that had been crushed under the weight of the political-philosophical preponderance of liberalism in the liberal-democratic conception. First and foremost in this front was the task of restoring to democracy its moral substance, which liberal democracy had squeezed out. This task was spearheaded by Macpherson's retrieval project. Macpherson argued that democracy constituted "a set of moral ends."[17] The efforts to retrieve the moral content of democracy went hand in hand with criticizing the liberal-democratic conception of democracy that had reduced democracy to a mere procedure or method for decision-making. Both Pateman and Macpherson attacked Schumpeter's conception for taking democracy as a procedure paralleling the competitive process of the free-market economy. In explaining the reasons for the perversion of the idea of democracy, Macpherson argued that the liberal state adopted democracy as a self-reformatory measure in order to diffuse the discontents of the working and poor classes, and thus as a means to prevent working-class revolutions.[18] Moreover, Macpherson argued that the "liberal democracy . . . was brought into being to serve the needs of the competitive market society." [19]

In his struggle to retrieve the moral content of democracy, Macpherson argued against the "market morality" and the market-based and "possessive" conceptions of Man that posited him as an "infinite appropriator"—a being who constantly gives in to his insatiable and incessant desire to appropriate more and more.[20] For Macpherson, a moral conception of democracy was intrinsically bound with an "ethical" or "developmental" conception of Man.[21] Macpherson conceptualized Man as a maximizer of his powers, as well as a "developer," and "enjoyer," and an "exerter" of his human capacities.[22] Man exerts his powers in order to realize his human potentials and to develop them further. It is through this exertion and its outcomes that he finds his fulfillment and enjoyment. In one sense, for Macpherson, the project of retrieving the moral content of democracy is essentially the problem of getting the "ontological" view of the human "essence" right—i.e., holding on to the ethical conceptions of Man and discarding the possessive ones.[23] If this can be done successfully, it would then follow that only a moral conception of democracy would be capable of truly accommodating the ethical conception of Man.

But, what exactly constituted the moral content of democracy? By all indications, it seems that both Macpherson and Pateman understood this content primarily as *substantive (i.e., economic) equality* and secondarily as *collective sover-*

eignty. One should add that the latter concept for both authors was understood as the right and freedom of the individual to exercise control over the societal decision-makings that affect her life. In the case of Pateman, the emphasis on direct participation was stronger than Macpherson's. Following Rousseau, Pateman emphasized the educative utility of participation in public and political matters.

*   *   *   *   *

The attacks of Macpherson and Pateman and like-minded thinkers on the foundations of liberal democracy were often interpreted as direct attacks on the individual liberties (and thus were regarded as inimical to individual rights) by the defenders of the status quo who enjoyed the luxury of preeminence and ubiquity in elite intellectual and media circles.[24] What made the idea of participatory democracy appear hostile to rights was that both Macpherson and Pateman focused their attention on democratic participation and control at the level of economic activities and resources, and thus targeted liberalism right where it mattered the most, viz., its attachment to property rights. As a result of being regarded as inimical to rights and liberties, the idea of participatory democracy did not find many receptive ears. What made the matter worse for the idea was the shift to the right in the 1980s, and the general retreat and weakening of the workers' movements and unions. These developments rendered the ideas of practicing participatory democracy at the workplace and the people's democratic control over the economy appear more "utopian" than they had in the earlier years. Finally, given the fact that the theories of participatory democracy did not produce coherent and elaborate institutional designs or models for the organizational forms of the citizens' participation at macro levels, by the late 1980s, the whole project of participatory democracy appeared to mainstream political thought as a utopian notion that could not be taken seriously, and thus had to be abandoned to the trash bin of history.

Despite these unfavorable conditions, the idea of participatory democracy did not die out completely. Its kernel survived and continued to assert its presence in the literature that was committed to strengthening the position of democracy in the liberal-democratic formula. A good example here is Jane Mansbridge's *Beyond Adversary Democracy* published in 1980. Mansbridge spent the late 1960s and the 1970s conducting research and cases studies on the practice of participatory democracy in small venues in the United States. At theoretical levels, Mansbridge put forth a unique form of participatory democracy, which she called "unitary democracy," and contrasted it with "adversary democracy." Unitary democracy was characterized by its approach to resolving conflicting private interests within the context of common interests—which she regarded as primary—and also by its moral commitment to the equal respect for all. On the question of making collective decisions, unitary democracy was premised on seeking consensus in "face-to-face assemblies."[25] By adversary democracy, Mansbridge had in mind the prevailing practice of democracy in the U.S., which she argued, worked with the underlying assumption that interests were essentially conflict-ridden. Adversary democracy took the equal protection of conflicting interests as its moral maxim, and used the majority rule and secret ballots as its approach to the question of collective decision-making.[26]

Mansbridge's unitary democracy had more in common with Pateman's version of participatory democracy than with Macpherson's, in that it placed its emphasis on the educational benefits the participation could render to citizens in helping them to understand their common interests. Unitary democracy also underscored the equality of power in participation, as Mansbridge focused on examining the issues involved in face-to-face interactions at the micro levels of the workplace and the small communities.[27] Attention to these questions set unitary democracy apart from other theories of participatory democracy. In these respects, one should consider unitary democracy as the forerunner of the theories of deliberative democracy that would gain currency and respect in the ensuing decade.

Another good example here is Benjamin Barber's *Strong Democracy* published in 1984. As characterized by the author himself, the theory of Strong Democracy is "a distinctively modern form of participatory democracy."[28] Compared with Macpherson's emphasis on the democratic control of the economy, and Pateman's emphasis on democracy at the workplace, as the foci of participation, what is different and distinctive about Strong Democracy is that it takes *civic activism* or "civic revitalization" as its main area of concern. Strong Democracy, as Barber puts it, "places politics before economics and suggests that only through civic revitalization can we hope, eventually, for greater economic democracy."[29] As a consequence, compared with the earlier generation of participatory theories of democracy, a strong emphasis on substantive equality is absent in Strong Democracy.[30] Strong Democracy, using Barber's own words, should be characterized as "not government by all the people all of the time over all public matters, but government by all of the people some of the time over some public matters."[31] The theory, as this characterization suggests, intends to place itself somewhere between liberal democracy, on the one hand, and direct or "unitary" democracy (as Barber calls it), on the other hand.

The other distinctive feature of Strong Democracy is that, in some ways, it finds more in common with the theories of deliberative democracy—which will be discussed in the following chapter—than with the preceding generation of the theories of participatory democracy. That is to say, Strong Democracy is more of a form of *politics of dialogue* than a kind of *politics of direct participation*. A good way to describe this feature of Strong Democracy is to compare it with liberal democracy as Barber himself does. Unlike liberal democracy, which "understands politics as a means of eliminating conflict . . . repressing it . . . or tolerating it," Strong Democracy seeks to "*transform* conflict through a politics of distinctive inventiveness and discovery. It seeks to create a public *language* that will help reformulate private interests in terms [that are] susceptible to public accommodation."[32] As Barber himself puts it, "[t]he stress on *transformation* is at the heart of strong democracy."[33] It is this distinctive feature of Strong Democracy that makes it more of a deliberative type of theory in nature than participatory democracy.

What is truly strong in Strong Democracy is the attention the theory pays to the need to produce an institutional framework and "a systematic program of institutional reform" for realizing the idea, as well as suggesting a political program and political strategy for achieving the desired institutional reforms.[34] Barber proposes a twelve-point program that combines some time-tested ideas such as election by lot, referenda, and national civic education, with new ideas such as "electronic town meeting," "electronic balloting," and "multichoice" and "two-stage" voting.[35] The strengths aside, one main shortcoming of Strong Democracy is that it focuses on "local citizenship" and thus comes up short in matters of "national citizenship." Of the twelve points in the program, only two are concerned with the macro aspects of political participation. As Barber concedes, the focus of the theory on localism has the "defect of parochialism."[36] The other problem with Strong Democracy is that the scope and degree of local participation proposed by Barber in his program is too extensive, and thus could be regarded by many citizens as being too taxing on their limited time and energy resources. Finally, too much emphasis placed by Strong Democracy on face-to-face interactions, as will be argued in the following chapter, could prove to be counterproductive to the cause of furthering democracy. The common feature that the theory of Strong Democracy shares with theories of deliberative democracy leaves it vulnerable to some of the criticisms that are often directed against the theories of deliberative democracy. These criticisms will be discussed in some detail in Chapter 13.

Notes

1. Held (1987), p.257.
2. Gould (1988), p.21.
3. Pateman (1970), p.37.
4. Ibid.
5. Elster (1986), p.103.
6. Gould (1988), p.20, italics added.
7. Ibid., italics added.
8. Ibid.
9. Lindsay (1996), p.89, italics added. Lindsay quotes Macpherson (1977), p.103.
10. Ibid. Lindsay quotes Macpherson (1977), p.111, italics added.
11. This proposal of Macpherson for reforming the internal structures of the parties should be viewed against the backdrop of his argument that, in attempting to appeal to a mass electorate, political parties in liberal democracy have become "less responsible to the electorate" (Macpherson 1977, p.67). Moreover, as was mentioned in note 38 in Chapter 11, Macpherson argued that mass parties had become less democratic, for "[e]ffective organization required centrally controlled party machines" (ibid., p.68). Held interprets Macpherson's proposal for democratizing the internal structures of the existing parties as meaning that "[t]he party system itself . . . be re-organized on less hierarchical principles, making political administrators and managers more accountable to the personnel of the organizations they represent. A substantial basis would be created for participatory democracy if parties were democratized according to the principles and procedures

of direct democracy" (Held 1987, p.258). Finally, as was stated above, the goal of reforming the party system was part and parcel of the democratic movement of the 1960s and early 1970s. Two strategies proved more successful than others: (a) to pressure parties to hold popular primaries for party cadres and party nominees, and (b) to alter the internal balance of the power and leadership of the parties by organizing mass voter registrations and mass membership drives.

12. As Held notes, for Macpherson this meant "combining competitive parties and organizations of direct democracy" (Held 1987, p.258).

13. Macpherson (1977), p.113. Macpherson's pyramidal model also seemed to imitate the model of councils in the Soviet Union. (See Cunningham 2002, p.137).

14. Townshend explains that Macpherson "was . . . reluctant to specify his model in a detailed fashion on the grounds that this would inhibit the freedom of those operating political and economic institutions to work out their own destiny" (Townshend 2001, p.148).

15. Macpherson (1977), ibid., p.97, italics added.

16. Lindsay (1996), p.89, Macpherson (1977), p.103. One should add that Macpherson, as Flower has stressed, "was not . . . an assertive disciple of community" (Flower 1991, p.59). According to Flower, "nowhere does he [Macpherson] really concentrate on community. And Macpherson's assurances to the contrary, merely making a democracy into 'an equal human society' does not address those whose major goal is community. What it does do, though, is reaffirm what Macpherson is all about: economic equality" (ibid.)

17. Ibid., p.78.

18. This point was discussed in Part II. See also Macpherson (1973), p.6, and Macpherson (1965), pp.9-11.

19. Macpherson (1965), p.35.

20. Macpherson (1973), pp.34-35.

21. Ibid., p.9, p.40, p.52.

22. Ibid., pp.32-37.

23. For Macpherson, this meant seeing "man's essence . . . not [as] maximization of his utilities but [as] maximization of his human powers" (ibid., p.32). "Or we could say that man is neither an infinite consumer nor an infinite appropriator but an infinite developer of his human attributes" (ibid.)

24. This was in part due to the fact that "[i]n participatory-democratic literature," as Cunningham puts it, "very little attention . . . [was] . . . devoted to the topic of rights" (Cunningham 2002, p.134). This lack of attention to the question of rights was not an oversight, but a necessary consequence of privileging the democratic pole of the liberal-democratic conception at the expense of its liberal pole in this literature.

25. Mansbridge emphasized the benefits of face-to-face interactions. "Face-to-face communication," according to Mansbridge, "is . . . likely to increase accuracy of perception," and can "lead citizens to take a degree of responsibility for their vote that they do not take if they are not physically present at the time of the decision" (Mansbridge 1980, pp.272-73).

26. See ibid, pp.vii-xiii and pp.3-33 for some discussions on the differences between unitary democracy and adversary democracy.

27. Parts II and III of Mansbridge's work are devoted to analyzing and addressing the questions related to the fear of conflicts, consensus building, and resolving conflicting interests at the micro levels, ibid.

28. Barber (1984), p.117.

29. Ibid., p.305.

30. It is worth noting that, for Barber, the moral content of democracy is not a pre-given, and as a result, much depends on discourse and human relations. "Politics does not rest on justice and freedom; it is what makes them possible . . . politics is not the application of Truth to the problem of human relations but the application of human relations to the problem of truth" (Barber 1984, pp.64-65). On the question of the "link between philosophy and politics," Lindsay makes an interesting comparison between Barber's Strong Democracy and Macpherson's participatory democracy. While Barber's takes all truth as following from human relations, and thus regards the question of politics as "open-ended," Macpherson starts from a set of pre-given normative ends (which have been inferred from theories of human nature) and views politics as the process of achieving them (Lindsay (1996), p.96).

31. Barber (1998), p.74.

32. Barber (1984), p.119, italics added.

33. Ibid., italics added.

34. Ibid., p.263. The other strong feature of Strong Democracy is that it comes with its own philosophical-methodological core. Strong Democracy challenges the "Cartesian epistemology of liberal democracy" and proposes an "epistemology of process" (Barber 1984, pp.64-65).

35. Ibid., pp.267-307. A summary of the program is presented on p.307.

36. Ibid., p.302. He believes that his "universal citizen service" program overcomes this defect.

Chapter 13

Theories of Deliberative Democracy

The idea of deliberative democracy is relatively a newcomer to the scene of alternative approaches to the question of democracy. David Held's 1987 comprehensive study of democracy, *Models of Democracy*, is virtually devoid of any references to the phrase or the idea represented by it. By all indications, the phrase "deliberative democracy" was first coined in 1980.[1] However, the idea itself did not gain currency until the early to mid 1990s, when numerous books and articles appeared on the subject. Although the theories of deliberative democracy put forth in these works spring from a variety of philosophical and political traditions, and use numerous concepts and differing linguistic constructs, they all start with the conviction that public dialogue and public deliberations ought to be regarded as an essential component of democracy. Alternately stated, the fundamental idea underlying theories of deliberative democracy is the conviction that democracy derives its legitimacy from the participation of the citizens in public deliberation on the affairs of society. As Joshua Cohen, one of the original proponents of the idea, puts it,

> [t]he notion of a deliberative democracy is rooted in the intuitive ideal of a democratic association in which the justification of the terms and conditions of association proceeds throughout public argument and reasoning among equal citizens. Citizens in such an order share a commitment to the resolution of problems of collective choice through public reasoning, and regard their basic institutions as legitimate insofar as they establish the framework for free public deliberation.[2]

By and large, theories of deliberative democracy appropriate the essential components of the theories of participatory democracy as their foundational elements. Among these, one should mention the notion of the participation of individual citizens in politics and their contribution to decision-making, as well as the critique of the conceptions of the individual often attributed to liberalism. However, in direct contrast to the theories of participatory democracy, broadly speaking, theories of deliberative democracy are keener on separating the political sphere from the economic realm than the theories of participatory democracy were. What distinguishes these theories from the original theories of participation is that they focus the attention of the theory on one particular aspect of the participation that escaped the attention of the early theorists, viz., the actual process of decision-making at micro levels.[3] Implicit in the theories of participatory democracy was the idea that the decision-making process would require deliberation. It was assumed that decisions would be arrived at by deliberative processes in which diverse views would be discussed and eventually *transformed* into some sort of convergent position.

It is this process itself that constitutes the point of departure for most theories of deliberative democracy. The conceptualization of democracy as public deliberation in these theories envisions the practice of democracy in terms of a constellation of social-dialectical processes that attempt to *transform* individuals' (often narrow and selfish) preferences and (self-interested) opinions to some sort of morally and rationally justifiable and convergent positions. The driving forces behind achieving these transformations or arriving at convergent positions are three-fold. The first is the power of reason; second, the participants' (moral) "commitment" to allow themselves to be persuaded by this power; and finally, the considerations of the common good.[4] The notion of decision-making in deliberative democracy is understood primarily as a matter of *consensus building*. The aim is to arrive at some consensus (but not necessarily unanimity). No one is expected to change her original opinion or be indifferent to her particular interest. What counts is that participants in deliberation are expected to seek consensus. That is to say, they are expected to allow themselves to be persuaded by the "force . . . of the better arguments" offered to them by their fellow deliberators.[5] The following is an account of how David Miller describes a deliberation session idealized in his conception of "politics as dialogue":

> A dialogue develops within which people may be led to revise their original opinions radically. Arguments are trumped by better arguments, until eventually a consensus emerges. What shape this consensus may have . . . varies according to the precise conception being employed. The important point is that people's adherence to the consensus view does not depend on its proximity to their original opinion, nor on strategic considerations, but on the strength of the arguments that have been offered for it. Once consensus is reached, it is formally adopted as common policy.[6]

The conception of decision-making idealized in deliberative democracy is often contrasted with the general understanding of decision-making in the liberal-democratic enterprise that, broadly speaking, takes democratic decision-making as a matter of procedural undertaking intended for reaching optimally "fair" *compromises* among different or opposing pre-figured interests.[7] (Recalling from Part II, this was referred to as the liberal-democratic "principle of collective-procedural decision-making.") In deliberative democracy, public deliberations are assumed to "be organized around appeals to the common good," and within the framework of some generally or commonly accepted moral and rational principles. As to identifying a framework for deliberating "rationally," one can turn to Joshua Cohen's notion of "politically acceptable reasons."[8] Not every reason is politically acceptable. "The background conception of citizens as equals," as Cohen argues, "sets limits on permissible reasons that can figure within the deliberative process."[9]

In some theories, deliberation is understood as a "cooperative search for truth," and in a sense, also as a matter of "rational political will-formation."[10] A decision made in this process is an "agreed judgment," a "collective verdict," and a product of "free and reasoned agreement among equals."[11] Moreover, a decision made deliberatively represents "pooling of judgements," a "collective judgement of reason," i.e., considered and reflected-upon judgements.[12] Still more, such a decision can be regarded as reflecting the "*generalizable* interests" as opposed to being a mere compromise reached among opposing particular interests.[13]

The language of deliberative democracy is at times opaque and incomprehensible. Habermas', Cohen's, and Gutmann and Thompson's versions are three cases in point. The language suffers from both over-theorizing and the lack of clarity or the inability to descend from its abstract theoretical framework, and propose practical models or produce concepts that could lay the grounds for conceptual model-making or institutional designs. In addition, most theories of deliberative democracy are presented in general terms and do not specify whether they are addressing the "macro" or "micro" levels. Nonetheless, given that deliberation is often understood as a face-to-face interaction, one can assume that deliberative conceptions are intended, or rather are best suited, for micro levels.[14]

Habermas' theory is a good case in point. Habermas does not make it clear which levels and structures of political power he is addressing—although one suspects that he primarily has micro levels, and perhaps also the "meso" levels, in mind.[15] As to the macro levels, it seems that Habermas proposes the creation of a set of associations (free and secondary to the state) in civil society (the public sphere) which would be autonomous from the state—(and one should assume that they would also be autonomous from the market)—and would involve citizens in opinion- and will-formation using democratic procedures in ideal speech-situations. (One should also assume that these institutions are intended to educate the citizens.) At the end of their deliberations, these associations would make decisions. However, the decisions rendered in the process would be non-binding, and thus would have only the status of recommendations. It is hoped that the elected officials would reflect on these recommendations as they engage in legislating laws and making public policies.[16] Habermas speculates that the fear of being voted out of office would compel the elected officials to heed the recommendations made by the civic associations.[17] By all indications, in Habermas' version of deliberative democracy, the civic associations are intended as *indirect* means for influencing the legislative and policy-making process. What is more is that Habermas' conception is developed with an eye on addressing the question of the sovereignty of the people. Interestingly, the conception addresses the question by sublimating it. As Habermas himself puts it, the conception "sublimates," "desubstantializes," and "proceduralizes" and "disperses" sovereignty and replaces it with the "*communicative power*" of the people expressed in the recommendations they make to the elected officials.[18]

While Habermas' emphasizes the epistemic aspect of deliberation, Cohen's version of deliberative democracy puts the spotlight on the moral (i.e., the egalitarian) aspect of the legitimacy that deliberation can bestow on democratic political institutions.[19] Cohen's deliberative democracy is essentially an ideal proceduralist version of democratic legitimacy that is based on the three principles of "deliberative inclusion," "common good," and "participation."[20] The first principle guarantees the basic liberties of all. The second demands that the interests of all should be the primary focus of decision-making. Finally, the third principle is intended to provide "equal opportunities for effective [political] influence" for all citizens.[21] These three principles, Cohen claims, constitute the "substance" of his proceduralist version of deliberative democracy.[22] In Cohen's account, deliberative democracy takes associations as the primary context for practicing the idea.[23] Cohen suggests the formation of "new arenas for public deliberation that lie outside the conventional political arenas" (e.g., outside the party and the state) that would create "new bases of social solidarity *through* a process of defining and addressing common concerns."[24] Moreover, Cohen seems to be more emphatic than Habermas on the pluralistic implications of deliberative democracy.[25] All in all, for Cohen, deliberative democracy is more than just a form of politics. It is rather "a framework for social and institutional conditions that facilitate free discussion among equal citizens . . . and ties the authorization to exercise public power . . . to such discussion."[26] Associations in Cohen's theory furnish the basic structures for this framework.

Another approach to deliberative democracy that justifies the idea in terms of moral legitimacy is Gutmann and Thompson's. In one sense, their view can be characterized as presenting deliberative democracy as a "moral" solution to the "problem of moral disagreement" that they regard as the most formidable challenge that faces American liberal democracy.[27] "Deliberation contributes to the legitimacy of decision made under conditions of scarcity," that is, in circumstances where there are numerous conflicting moral positions and the democratic government has to render a single decision.[28] The deliberative way of making decisions is to seek a "morally justified consensus."[29] Morally justified consensi build up in part through offering moral justifications for supporting the decisions that deliberative democracy makes. Moral justifications "help sustain political legitimacy," which in turn secures the future of the establishment and encourages citizens to "live with one another civilly."[30] Moreover, moral justifications may not always satisfy everyone or resolve moral disagreements, and this is why we need to turn to the "principles of accommodation" that are based on "mutual respect"—the same value that lies at the core of the principle of reciprocity, and the idea of deliberation itself.[31] "Mutual respect not only helps sustain moral community in the face of conflict but also can contribute toward resolving the conflict."[32]

Carlos Santiago Nino's conception of deliberative democracy, on the other hand, claims to take into account both the epistemic and moral aspects of the idea of deliberative democracy. His "epistemic theory of democracy," as he puts it, is "deeply intertwined with morality" and takes deliberative democracy as a procedure for "transform[ing] people's preferences into morally acceptable ones."[33] In these essential aspects, Nino's theory parallels Habermas'—notwithstanding his insistence that his ontology and epistemology of social morality is superior to Habermas' and that his is the "most reliable" method of reaching morally acceptable solutions.[34] Nino takes great care not to lose sight of the importance of the liberal assumptions and values that are deeply entrenched in modern liberal democracies.[35]

As to laying theoretical grounds for model-making, Nino is more explicit in providing a vision of how deliberative democracy ought to function, and in this respect he fares far better than most other theorists. This is vividly evident in his discussion of the question of representation. He regards representation as a necessary evil. For Nino, practical considerations and "time limit" necessitate the "passage from desire for unanimity to majority rule."[36] The same considerations also necessitate the passage from the desire to have decisions made by the citizens to having them made by their representatives on their behalf. Nino conceives of representation as a "delegation to continue the discussions from the point reached by the electors during the debate leading to the choice of representatives."[37] Finally, just as Habermas', Nino's theory is also keen on dispersing sovereignty, but primarily as a measure for preventing the monopolization of the power by interest groups or factions.[38] However, unlike Habermas, Nino articulates in precise language what exactly he means by dispersing the will of the people. Stated summarily, Nino envisions the dispersion of sovereignty into various representative bodies on the local, state, and national levels, in very much the same way he believes the model of federalism functions in the United States.[39]

Another theory that focuses on the macro level is James S. Fishkin's theory of "deliberative opinion polling" presented in his *Democracy and Deliberation*. In this theory, deliberation takes place among a representative sample of citizenry, a "statistical microcosm of the society," about 600 individuals—and it seems that the theory suggests that this sample be selected by lottery.[40] This representative sample is to be brought together in a convention ("National Issues Convention") for a specific number of days in order to deliberate (argue, discuss, debate) in face-to-face and small group settings, which could be nationally televised.[41] This convention could be formed early in the presidential primary campaign to discuss what the nation expects of the next president and to define the presidential campaign issues for the candidates. It is also possible to form conventions of a similar type on the major issues of concern to the nation. The main idea here is that deliberation would ensure that opinions formed by this method would be intelligent ones. Once the deliberation is completed, then the sample will be polled and the results will be publicized as an indication of how the nation views the issue at hand. The results will also be presented to the citizens and the politicians and elected representatives as indications of where the nation stands on the issues and what they believe should be done about them. The role of deliberative opinion polls will be "advisory" in nature.[42] Nonetheless, the opinions produced by polling are expected to be taken by the politicians as a bedrock or a framework for legislative and policy-making purposes.

*   *   *   *   *

This brief survey of the theories of deliberative democracy, and the discussion of some of the themes they preoccupy themselves with, points in the direction of identifying three categories of problems that hound these theories. The first category brings to light some of the difficulties present in the parameters and requirements that establish the framework for dialogues in deliberation sessions. The second category includes a number of problems that question the feasibility of practicing deliberative democracy. Finally, in the third category, one finds a set of questions as to what ultimate ends the theories of deliberative democracy pursue, as well as some questions as to what sort of claims one should consider as forming the underlying premises of the idea of deliberative democracy. In what follows, these three categories of problems will be briefly discussed.

As to the problems in the first category, the underlying difficulty stems from requiring (public) reason, and the citizens' commitment to abide by its dictates, as the primary, if not the only, determinants in reaching decisions in deliberation sessions. In very much the same way that Communitarians overburden the populace morally with their expectations of virtue, Habermas and Cohen, and to a lesser degree Nino, overburden citizens with demands of rationality (or reasonableness). Cohen requires that the members should have "deliberative capacities, i.e., the capacity . . . for entering into a public exchange of reasons and for acting on the result of such public reasoning."[43] Moreover, Habermas and Nino also overburden citizens with some demands of morality. Another group of problems in this category springs from the unrealistic (implicit) assumption that the citizens' conceptions of public interests, as well as their conceptions of how these interests could be furthered, are subjectively constituted, and thus could be altered or transformed by educating the citizens on social facts and truths, as well as by reasoning, *in the course of dialogue.* Moreover, another set of problems here arises from a similar type of assumption regarding the nature of private interests, as well as the further assumption (and the requirement) that these interests could (and should) be altered when they cannot stand up to the scrutiny of reason. Such assumptions ignore the simple fact that some individuals—be it for their own private interests or because of their personal convictions or their ideological commitments to some belief system—would never be persuaded by any sort of reasoned and morally based arguments on some issues. No doubt, subjecting these individuals to public reasoning on these matters and expecting them to give up their views would be oppressive, as well as being counter-productive to the cause of democracy.

As to problems with the feasibility of practicing the idea of deliberative democracy (the second category of problems), the main difficulty with implementing the idea at macro levels (involving over 10,000 individuals), or even at meso levels (involving around 5,000 individuals or less), has to do with questions of practicality and efficiency. In considering the tremendous logistic difficulties involved in assembling the citizens at macro levels, or even at meso levels, as well as in reflecting upon the length of time it would take to build consensus via interpersonal dialogues in such assemblies, one cannot help but think that the idea of practicing democracy in deliberative modes at macro levels (or even at meso levels) would be a highly impractical and inefficient, if not an impossible, way of making public decisions. The other feasibility problem that could be encountered in practicing democracy in deliberative ways at macro and meso levels can be stated as follows: the familiar difficulties associated with working within large crowds and the time-consuming features of deliberations, and also the strict set of rules and requirements prescribed for deliberation, would discourage most citizens from attending macro or meso assemblies. In such assemblies, one should consider the high probability that only a small minority of citizens (the most zealous individuals and the most ardent members of the special interest groups) would attend these deliberation sessions. Moreover, in large crowds, only a small number of the individuals would have the opportunity to speak. In view of these difficulties, one would suspect that the large majority of citizens would elect to stay out of the deliberation sessions.

A good example of this latter problem at macro or meso levels would be the actual experience of the Athenian Assembly meetings. The quorum for Assembly was 6,000, which reflected the maximum capacity of the amphitheater. Despite the fact that the citizens were paid to attend the Assembly (starting from 404-02 B.C.), the quorum was not achieved regularly; the actual attendance was "much less." And this was in the face of the fact that the size of the eligible citizen body during this period is estimated in the range of 20,000-40,000.[44] Finally, one should always be wary of the danger that lurks behind every large citizens assembly; namely, these assemblies are vulnerable to being hijacked by charismatic public speakers and would-be demagogues. It was this aspect of the Athenian democracy that moved Rousseau to claim that "Athens was in fact not a democracy, but a very tyrannical Aristocracy, governed by philosophers and orators."[45]

Having eliminated the macro and meso levels for being too large of a venue for practicing deliberative democracy for feasibility and efficiency reasons, the only arena that seems to hold a promise for deliberative democracy is the micro level (10 to100 individuals). Here, having eliminated the logistical problems of organizing the deliberation sessions and the problems of functioning in large crowds, one creates a venue which is more suitable for interpersonal dialogues. However, even here the idea of deliberative democracy runs into difficulties. In addition to grappling with the old and familiar problems that beset interactions within groups, the theories of deliberative democracy bring into the sessions a new set of problems by expecting that a decision be reached at the conclusion of the session, and that this decision be arrived at via reaching a consensus. One should recall that the consensus in question is expected to be reached within the strict framework of requirements and parameters that these theories prescribe. This framework demands reasonableness and requires the commitment that the participants change their views when they are presented with a better argument.

One should question whether this scenario idealized by the theories of deliberative democracy could ever be closely approximated at micro levels. To begin with, given this strict framework, most group members or would-be participants might be discouraged from participating in deliberation sessions because the deliberation process could take too long or could be hijacked by compulsive talkers and "orators." Another discouraging factor here would be that the sessions could be too argumentative and stressful for most participants—for they have to sit through and witness the zealous speakers, opinionated individuals, and adamant supporters of the contending views fight out their battles in front of the participants. "Face-to-face assemblies designed to produce feelings of community," as Mansbridge concedes, "can thus backfire . . . the fear of open hostility is an important cause of nonparticipation in politics."[46] Given these dangers, the micro level deliberation sessions could become divisive and tear a community of people apart if they are not managed well. Moreover, micro deliberation sessions would be avoided by introverted individuals, and could be oppressive for them if they are required to sit in the meetings. The deliberation sessions, given their framework of requirements and parameters, no doubt, would favor those who are extroverted or have the mastery of the art of argumentation and the necessary skills to conduct "formal debates," and thus can argue their points effectively as well as dispassionately.[47] In this sense, deliberation sessions will belong to those who have the necessary intellectual-dialectical resources to make their reason prevail.[48] Furthermore, the pressure of reaching a consensus, as the expected outcome, could warp the views expressed and would force some to give in to the opposing arguments unwillingly and without being convinced. "Consensus, when it emerges," as Femia argues, "may be due to conformity rather than rational agreement. People in collective settings appear only too ready to conform to the majority in the group and to abandon their own personal beliefs and opinions."[49]

The problems of feasibility and efficiency associated with practicing the idea of deliberative democracy at the macro levels are perhaps the main reasons why theories of deliberative democracy shy away from proposing institutional frameworks or designs for realizing the idea at the macro levels. Broadly speaking, and with the possible exception of Fishkin's and Habermas', theories of deliberative democracy appear to be content with expressing the idea of deliberative democracy only in abstract propositions such as "government by discussion" or "interaction . . . [constituting] . . . the source of democratic legitimacy," or "mak[ing] democracy more deliberative."[50] In the case of Habermas and Fishkin, although they manage to descend to concrete levels, the institutional frameworks or models they propose, as will be seen below, raise more questions than they answer. As was stated earlier, Fishkin's proposed solution to the problems encountered at the macro levels was to elect—by lot—a representative sample of the people who would be commissioned to do the deliberating on behalf of the entire populace. Given that the representative deliberators represent a statistical sample of the nation, Fishkin argues that the results obtained by the deliberation of the representative deliberators could be regarded as representing the views of the entire nation. Although Fishkin's solution seems to solve the practicality and efficiency problems, it gives rise to another set of questions that go to the very heart of the idea of deliberative democracy. These questions fall within the third category.

*   *   *   *   *

Now, a discussion of the third category of problems. The questions raised by Fishkin's solution discussed above can be posed as follows. What is of greater importance, or of more fundamental value, to theories of deliberative democracy: the idea that citizens ought to *contribute* to political decision-making, or the idea that they ought to *participate* in public deliberations? Moreover, what is *the* main fundamental idea underlying the concept of deliberative democracy: citizens' contribution or their public deliberation? One problem with Fishkin's solution is that it comes at the cost of betraying the idea of *public* deliberation and the citizens' *participation* in it. It is not the public itself, but its representative sample which gets to deliberate. The other problem with Fishkin's solution is that it does not fully embrace the idea of the *contribution* of citizens to political decision-making, for the results produced by the deliberation of the statistical sample are assumed to be *advisory* in status, thus having no legally binding character, or no assurance that they would be considered in political decision-making. One should add that this latter shortcoming also hounds Habermas' proposed model for the macro levels, albeit his model affirms the value of the citizens' *participation* in public deliberations.

In view of this discussion of Fishkin's solution, as well as in light of the survey of the theories of deliberative democracy presented earlier, it appears as if the ultimate issue of concern for theories of deliberative democracy, at least at macro levels, is not the question of the citizens' *direct participation in decision-making*, nor their *actual contributions*, but that of their *indirect and potential contribution* to the process. This is evident in Habermas, Cohen, and Fishkin, and in a lesser degree in Nino. In light of these considerations, the theories of deliberative democracy, despite what they claim, fall short of offering a genuinely different sort of approach to the question of political decision-making at macro levels. Despite their sharp criticisms of liberal democracy on the question of public decision-making, theories of deliberative democracy conform, at their very core, to an essentially liberal-democratic principle. That is to say, in the deliberative scheme of things, despite the space carved up for public deliberation, public decisions and policies would continue to be made the old liberal-democratic way—i.e., *not by the citizens themselves, but by their elected officials and representatives* (virtual representatives in the case of Habermas and Cohen, and functional representatives in Nino's case). By all indications, the "agreed judgements" and "collective verdicts" reached by citizens in the deliberative associations or in public arenas would not have the status of binding decisions, laws, or policies; but rather that of suggestions and recommendations that would be delivered to the lawmakers (political elites, the representatives) to be taken into consideration at their discretion in making policies and laws.

In light of what has been presented thus far, one can argue that what the theories of deliberative democracy intend on accomplishing amounts to a considerable retreat from the original retrieval task that was spearheaded by the founding theorists of participatory democracy a generation earlier. One cannot but surmise that this retreat, in part, reveals the extent of the deformity the originally bold idea of participatory democracy has suffered under the weight of criticisms and suspicions that regarded it as hostile to the values and institutions of liberal democracy (e.g., civil liberties, protection of privacy, and separation of the political and the economic). Deliberative democracy manages to escape these criticisms or suspicions in part by keeping citizens at a distance.[51] Deliberative democracy succeeds in this by deforming the originally bold idea of citizens' *actual* participation in politics into that of their *potential and indirect contribution* to it.

The retreat has led the retrieval project back to the liberal-democratic conception, and thus back to "procedure." The original retrieval task of substituting "substance" for "procedure," or privileging it over procedure, has been contorted into the goal of breathing substance (i.e., fairness and equal opportunity) into procedure. As Richardo Blaug seems to hint, the struggle to overthrow procedure has been revised to a struggle for "procedural fairness," i.e., the struggle for a truly *equal access to the process of influencing the representatives* or the lawmakers.[52] Deliberation is assumed by some theorists to be the right means for achieving this end.[53] As a consequence of this turnaround, theories of deliberative democracy have managed to fare much better than theories of participatory democracy in defending themselves against the liberal-democratic critics, and thus have succeeded in securing a place for themselves in the realm of the acceptable and respectable— or as Bohman likes to phrase it, they have "come of age."

By "coming of age," Bohman means to convey two important facts about the theories of deliberative democracy in the closing years of the 1990s. First, they "have come to emphasize the *process* of deliberation itself rather than its *ideals* and counterfactual [i.e., impractical and inefficient] conditions and procedures."[54] Second, they have increasingly preoccupied themselves with making the existing institutions of liberal democracy (e.g., voting, representation, constitutional law, the aggregative- and compromise-based conception of decision-making, etc.) "more deliberative than rejecting them for more direct democracy."[55] "This," as Bohman continues, "has led to an increasing emphasis on the epistemic as well as the moral aspects of *public justification*."[56] "Tempered with considerations of feasibility, disagreement and empirical limits," as Bohman approvingly reports, "deliberative democracy has now 'come of age' as a practical ideal."[57] The most important thing that has come out of this "coming of age," it seems, is that deliberative democracy has given up on demanding "too much social consensus or epistemic virtue," as it did in the late 1980s and early to mid 1990s.[58] The newfound "political realism" of deliberative democracy has led it to define its mission as, to quote Bohman, "to improve democratic practice in light of its morally inclusive, yet cognitively demanding ideal . . . [to provide] . . . an increasingly large space for politics within the normative constraints and the many empirically feasible variations of its ideal of *public reason*."[59]

More than anything, Bohman's notion of deliberative democracy's coming of age should be taken to signify that the theories of deliberative democracy have backed away from their earlier preoccupation with theorizing the idea of citizens' contribution to political decision-making—(be it direct and actual, or indirect and potential). In light of being awakened to problems outlined above, the new preoccupation of the theories of deliberative democracy seems to be with exploring some ideas that, if put into practice at macro levels, would amount to *not much more* than requiring the public officials or decision-making bodies to justify their views and decisions to the public on epistemic and moral grounds. That is to say, *the decisions to be made, or polices to be enacted, must be justifiable to the public* by appeals to some criteria of reason that take freedom and equality—and perhaps community—as their benchmarks. The underlying assumption here seems to be that if this requirement could be met satisfactorily, and within a fair procedural framework, then the resulting decisions or policies would be acceptable to all—at least in principle—and thus would be regarded as having democratic legitimacy.

Thus, by placing the idea of public justification at the center of the question of democracy, and by insisting that interests, views, and decisions be justified to the public by the "weight of character and strength of argument," deliberative democracy has come full circle to embrace a Burkean-Millian notion of deliberation *among the representatives*.[60] What is new in the late deliberative democracy is the added feature that deliberations, and thus public justifications, ought to take place in the *public sphere*. What is lost in this coming of age of deliberative democracy is the value assigned to citizens' actual and active participation in deliberation and thus their contribution to the decision-making process that constituted the hallmark of the theory in its earlier years—(albeit, even in those years, the value assigned to citizens' participation only implied that their contributions to decision-making would be indirect and potential).

A quick survey of some of the latest literature on deliberative democracy reveals that the theory continues to roam in pretty much the terrain carved out by the notion of *public justification in the public sphere* that heralded the maturation of the theory at the close of the 1990s. What seems to be coming close to making headway out of this terrain, as if by the force of the logic unleashed by the notion of "public justification" itself, is the idea of "public contestation." In light of the fact that the "public justification" version of deliberative democracy assigns no role for citizens' actual or direct contributions to the decision-making process (as is the case with the present system of governing), it is forced to validate its claim to representing a richer or more legitimate notion of democracy by showing its superiority to liberal democracy somewhere outside of the decision-making process. In substantiating this claim, nothing seems more reasonable than to argue that citizens ought to have the opportunity to "contest" the justifications offered to them by the decision-makers.

The shift from emphasizing the idea of citizens' participation in decision-making (albeit indirect) to underscoring the democratic merit of "contestations" is clearly present in John S. Dryzek's *Deliberative Democracy and Beyond: Liberals, Critics, and Contestations*—perhaps the most important piece of work on deliberative democracy so far in the new century. Dryzek argues for "detaching" the idea of deliberation away from the state (or "constitutional surface of political life" or "liberal constitutionalism," as he calls it, i.e., where the decisions are made) and planting it in the public sphere.[61] The central idea behind "contestations" in the public sphere is to transform people's opinions, i.e., to change the "balance of discourses" in the civil society in ways that would result in arriving at some substantive "public opinions" (the "provisional outcomes of contestations") that would then be "transmitted" to the decision-makers for consideration.[62] Consequently, the criterion of democratic legitimacy in Dryzek's version of deliberative democracy is the decision-makers' "responsiveness" to the contestations of their decisions by the public that would be "transmitted" to them.[63] In many respects, Dryzek's notion of deliberative democracy marks a return to Habermas' and Fishkin's notions of deliberative democracy discussed earlier in this chapter.[64]

Notes

1. According to David M. Estlund, "[t]he phrase appears to have been coined by Joseph M. Bessette, in his "Deliberative Democracy: The Majority Principle in Republican Government," in *How Democratic Is the Constitution?* (Robert A. Goldwin and William A. Schambra, eds., 1980)" (Estlund 1993, p.1437).

2. Cohen (1997a), p.72.

3. As was mentioned in Chapter 12, Mansbridge's unitary democracy was an exception in this case.

4. "While no one is indifferent to his/her own good, everyone also seeks to arrive at decisions that are acceptable to all who share the commitment to deliberation" (Cohen 1997b, p.75).

5. The phrase inside the quotation marks is borrowed from Habermas' *Legitimation Crisis* in a discussion presented in the chapter titled "The Relation of Practical Questions to Truth" (Habermas 1975, p.108).

6. David Miller (1989), p.255.

7. In the "liberal view," as Habermas puts it, "the democratic process is effected exclusively in the form of compromises among interests" (Habermas 1996, p.296).

8. Cohen (1997b), p.417.

9. Ibid., p.421.

10. This is the Habermas' version of deliberative democracy. The idea of the "cooperative search for truth" appears in his *Legitimation Crisis* in the chapter titled "The Relation of Practical Questions to Truth" (Habermas 1975, p.108). This idea is essentially a Rousseauean construct. As one can recall from Part I, the process of formulating the "general will," ideally speaking, is a process of discovering the truth—see Andrew Levine (1976) for a discussion of this point, e.g., p.56. Moreover, the notion of the "rational political will-formation" is also Habermas' and appears in Habermas (1996), p.59.

11. The first two characterizations are Femia's, the last Joshua Cohen's, quoted by Femia. (See Femia 1996, p.365.)

12. These characterizations are Michelman's, quoted by Estlund (1993), p.1442.

13. The phrase "*generalizable* interests" is borrowed from Habermas' *Legitimation Crisis* in a discussion he presents in the chapter titled "The Relation of Practical Questions to Truth" (Habermas 1975, p.108, original italics).

14. According to Blaug, the main area of deliberative democracy is the micro level, and even perhaps the meso levels. At macro or state levels, theories of deliberative democracy run into serious difficulty and "move away from the macro level" (Blaug 1996, p.56).

15. By "meso" levels, it is meant larger groups. Following Blaug's characterization, here the meso levels would include "civil associations, social movements, ethnic and religious groups, firms and the institutions of civil society," Blaug (1996), p.56.

16. Habermas (1996), pp.59-60.

17. Ibid., p.60.

18. Ibid., pp.58-59.

19. "The fundamental idea of democratic . . . political legitimacy is that the authorization to exercise state power must arise from the *collective decisions* of the *equal* members of a society who are governed by that power" (Cohen 1998, p.185, first italics original, second added).

20. Cohen (1997b), pp.414-23.

21. Ibid., p.422.

22. Ibid., p.431.

23. Ibid., pp.426-31.

24. Ibid, p.429, original italics.

25. Cohen (1997a), p.72.

26. Cohen (1997b), pp.412-13.

27. Gutmann and Thompson (1996), p.1. Gutmann and Thompson over-emphasize the importance of "moral disagreements" in the existing liberal democracy, as if all conflicts related to interests are rooted in *moral* disagreements among the contending parties. They present this question as the most important problematic of liberal democracy, thus leaving out its two main problems which were characterized earlier as problems posed by "fund democracy" and "audience democracy."

28. Ibid., p.41.

29. Ibid., p.42.

30. Ibid.

31. Ibid., p.79

32. Ibid., p.80.

33. Nino (1996), p.143.

34. Ibid., pp.107-13 and p.144.

35. Ibid., p.143.

36. Ibid., pp.118-19

37. Ibid., p.133. Nino discusses Burkean and Millian conceptions of representation in pp.146-47 and pp.171-73.

38. Ibid., p.166.

39. Ibid.

40. Fishkin (1991), p.93.

41. Ibid.

42. Ibid., p.95.

43. Cohen (1997a), p.73.

44. Budge (1996), p.25.

45. Quoted by Femia (1996), p.389.

46. Mansbridge (1980), p.273.

47. As Bohman reports, some critics of deliberative models, e.g., Lynn Sanders and Iris Young, "see deliberative models as overly cognitivist or rationalistic and thus insuffi-

ciently egalitarian: it favors the educated and the dispassionate and excludes the many ways that many people communicate reasons outside of argumentation and formal debate, such as testimony, rhetoric, symbolic disruptions, storytelling and cultural- and gender-specific styles of communications" (Bohman 1998, pp.409-10).

48. Or to use Mansbridge's words, "a face-to-face assembly lets those who have no trouble speaking in public defend their interests; it does not give the average citizen comparable protection" (Mansbridge 1980, p.274).

49. Femia (1996), p.385. The problems associated with face-to-face deliberations at the micro levels are further discussed in Chapters 3 and 4 of *Direct-Deliberative e-Democracy.*

50. The characterization of deliberative democracy as the "government by discussion" is Sunstein's (Sunstein 2001a, p.45). Taking "interaction" as the "source of democratic legitimacy" is the characterization used by Blaug (Blaug 1996, p.69). Finally, the last phrase quoted above was from Gutmann and Thompson (1996), p. 358. This is how Gutmann and Thompson evade the challenge of institutional design: "We did not undertake to provide an inventory of institutional changes because the design of the institutions of deliberative democracy depends critically on developing principles to assess them . . . Once the principles of deliberative democracy are better understood, the search for their most suitable institutional expression can become more productive. The best forum for considering the design of deliberative institutions is likely to be one in which deliberation, however nascent, has a prominent place" (ibid.).

51. As will be seen later in this chapter, deliberative democracy also manages to escape these criticisms by staying away from questions of substantive equalities and economic democracy.

52. Blaug (1996), p.53-54. Equal in the sense of the public being on the same footing as the traditional influence peddlers and lobbyists.

53. Blaug includes Barber and Habermas in this category (Blaug 1996, pp.51-52). Cohen also belongs to this category.

54. Bohman (1998), p.401, italics added.

55. Ibid.

56. Ibid., italics added.

57. Ibid., p.422.

58. Ibid.

59. Ibid., p.423, italics added.

60. The phrase inside the quotation marks is borrowed from Mill (1861), p.127. As was discussed in Part II, Madison's conception of representation, while it acknowledged the existence of diverse and varied interests in society, its main shortcoming was that it lacked "a politics of deliberation *over the common good*" (Williams 1998, p.42, italics added). For Madison, the guiding principle of deliberation in decision-making was to strike balances and reach compromises among competing interests. Burke's conception, on the other hand, while it failed to see the diversity of interests, and thus the need for compromise-making, its advantage over Madison's is that it placed the focus of deliberation in the representatives' assemblies *on the matters of common good.* And it is in this sense that the mature theories of deliberative democracy are characterized above as embracing a Burkean-type notion of deliberation. Finally, the Millian connections of the mature theories of deliberative democracy have to do with the fact that they have transferred the deliberative powers to the elites, and hence have sidelined ordinary citizens. Moreover, all theories of deliberative democracy can be characterized as Millian in that they follow in the footsteps of J. S. Mill in attempting to place democracy within a framework where the "interests, the opinions ...[are] heard, and . . . have a chance of ob-

taining by weight of character and strength of argument an influence" (Mill 1861, p.127). It should be noted that Mill advanced this perspective on democracy in opposition to the idea of the rule by the numerical majority (of the poor).

61. Dryzek (2000), e.g., p.175. Detaching the idea of deliberation from the state does not necessarily amount to backing away from the idea of "public justification." Knowing that their decisions could be contested, as well as being in tune with "public opinion," the elected public officials will make decisions with one eye on justificatory reasons that they might have to present to the public in support of their decisions.

62. Ibid., pp.49-51, also in Dryzek (2001), p.659 and p.663.

63. Or, as Dryzek himself puts it, "legitimacy can be sought . . . in the provisional outcome of the contestation of discourse in the public sphere as transmitted to the state or other authorities" (Dryzek 2001, p.666).

64. Dryzek's own view is that his is superior to Habermas' because "legitimacy for Habermas is secured by public acceptance of *procedural* responsiveness [of the lawmakers, and] not by the *actual* responsiveness of the process of legislation to the substance of public opinion on an issue" (Dryzek 2001, p.657, first italics original, second added).

Concluding Remarks to Part III

Amidst the talk of the "crisis of liberal democracy," the 1960s began with the hope that the ideal of "the establishment of a democracy of participation" could soon be realized.[1] Almost forty years later, as far as the eye can see, not only are there no visible signs of a "democracy of participation"; but worse, the best democratic hope for humanity seems to be nothing better than the idea of the establishment of a democracy of public justification that is currently being championed by theorists of deliberative democracy. Part III presented an overview of some of the problems that have dogged the "liberal-democratic conception of democracy" and the liberal-democratic state during the last four decades or so. Part III also sketched the theoretical contours of the rise and fall of the idea of "power to the people" (or "participatory democracy") that invigorated the enthusiasts of democracy in the period of the 1960-1970s. Finally, Part III presented a discussion of the idea of deliberative democracy that began in the 1990s and continued into the twenty-first century.

In many respects, the resurgence of the idea of "power to the people" or "participatory democracy" in the 1960s was the reincarnation of the original meaning of the idea of democracy. The best known theoretical representatives of this idea in the 1960s and 1970s in the North-American scene were Carole Pateman and C. B. Macpherson. Both Pateman and Macpherson were mainly concerned with the question of substantive equalities, which they believed had been squeezed out of the liberal-democratic conception of democracy. As to the question of sovereignty, both theorists viewed people's sovereignty as the right and freedom of ordinary citizens to participate in political decision-making. The focus of both theorists was on the question of democracy in the "economic society." While Macpherson was interested more in the question of the democratic control over the economy at the macro levels, Pateman seemed to be devoted more to the question of democracy at the workplace. Finally, placing this discussion of the theories of participatory democracy in the context of the two democratic shortcomings of liberal democracy identified in Part II, one can further argue that while Pateman showed more interest in addressing the first shortcoming (especially at the level of the workplace), Macpherson was devoted mainly to addressing the second difficulty.[2]

With the advent of the 1980s, the project of participatory democracy appeared to be doomed. However, in 1984, a "distinctively modern form of participatory democracy" resurfaced in Benjamin Barber's *Strong Democracy*. Unlike Pateman and Macpherson, Barber did not show much interest in the question of substantive equality or democratic control over economic activities or resources. Regarding the question of collective sovereignty, however, Barber fared better than the former authors. Strong Democracy proposed a system of initiatives and referenda at national levels that were intended to empower citizens to participate in the decision-making process at macro levels. The main shortcoming of the Strong-Democratic notion of sovereignty was that it did not take the exertion of collective sovereignty as an ongoing process, but rather as discrete events that would take place "some of the time." In other words, Strong Democracy understood democracy as the rule by the people, but only "some of the time." Thus, in this regard, Strong Democracy failed to fully elevate citizens to the position of the direct authors of the laws that they live under.[3]

As is the case with Strong Democracy, theories of deliberative democracy also, by and large, shun the question of democracy in the sphere of economics.[4] However, as was seen, unlike Strong Democracy's "hope" that political democracy would "eventually lead . . . [to a] . . . greater economic democracy," the latter theories seem to remain silent on the question of economic democracy.[5] As a consequence, theories of deliberative democracy do not show much interest in the question of substantive equalities, which Macpherson and Pateman took as constituting the main sense of the moral content of democracy. In theories of deliberative democracy, the only detectable sign of egalitarianism is the attention these theories pay to the question of fairness in procedures and their insistence that all political arguments presented by deliberators be treated equally (e.g., Cohen). On the question of sovereignty, on the other hand, things look far worse for theories of deliberative democracy. As is the case with liberal democracy, the idea that the people ought to be the direct authors of the laws they live under is completely sidetracked.[6]

In light of the discussion of the theories of deliberative democracy provided above, it becomes evident that, when compared with the liberal-democratic conception of democracy, there is nothing genuinely radical or different about the conception of democracy put forth by these theories. As is the case with liberal democracy, deliberative democracy in its matured state is *deeply committed to utilizing the existing form of representational government*. It would be safe to claim that the ultimate aim of deliberative democracy seems to be none other than (to use the phraseology of Gutmann and Thompson) to "make [the existing liberal] democracy more deliberative" and to help "satisfy the demand of democratic accountability" of representatives.[7] Alternately stated, the mission of deliberative democracy seems to be *to fill the presently existing "representation gap" between the citizenry and their representatives, and thus to offer a solution out of the current crisis of liberal democracy*.[8] Viewed from this angle, deliberative democracy should not be characterized as a variant of participatory democracy, nor as a distinct conception of democracy, but rather as a supplementary theory for the liberal-democratic conception.

Moreover, relating this critique of deliberative democracy to the discussion of the two shortcomings of liberal democracy identified in Part II (and their characterizations as "audience democracy" and "fund democracy" in the period of the last two decades as was done earlier in this part), one can venture into arguing that the enterprise of deliberative democracy could be characterized as a timid argument against "fund democracy." That is to say, the argument for "procedural fairness" (i.e., equal access to the process of influencing the representatives or public officials), as well as the argument that the decision-making bodies be subjected to public justification on epistemic and moral grounds, one can contend, are but philosophically camouflaged and politically concealed arguments against the undue power the lords of the "economic society" wield over the decision-making process in "fund democracy."[9] Thus, one suspects, the dismay that advancing direct arguments against "fund democracy," and consequently raising the bold idea of equal access to political power as a normative criterion, could be viewed as too political or distracting—(or perhaps that it could provoke fierce reactions or reprisals in the conservative atmosphere of the recent decades)—compels deliberative democrats to ignore the rampant popular discontent with "fund democracy," and instead present their case for democracy in the coded language of deliberative democracy.[10]

Even if one regards the case of deliberative democracy as a timid argument against "fund democracy," and counts its opposition to "fund democracy," however concealed, as its virtue, deliberative democracy still has to face the criticism that it suffers from the vice of elitism, and thus lacks a real solution to the problem of "audience democracy." No matter how one slices the idea of deliberative democracy, especially in its most recent manifestations, citizens come out as audiences. They still have no actual or direct role to play in decision-making. What appeared in the earlier theories of deliberative democracy as an active (albeit indirect and potential) role reserved for ordinary citizens to play, now seems to be taken away from them in the most recent versions of these theories. Ordinary citizens are no longer regarded as the main deliberators, but as the audience to deliberations that would take place among the elites. Hence, the show of "audience democracy" goes on, but this time in the newly built studio theatre of deliberative democracy. The main advantage of the new theatre is that the stage is lower and closer to the audience. This proximity makes the audience hear the elite actors better, and at the same time, makes the actors more conscious of the presence of the audience. Despite this advantage of the new theatre, the roles stay the same as the old theatre: actors remain actors and audiences remain audiences.

Notes

1. Quoted earlier from *The Port Huron Statement,* the founding document of the Students for a Democratic Society (SDS), in James Miller (1987), p.333.

2. The first of these shortcomings was that, in the liberal-democratic state, the representative form of government was elitist by design, and thus, was intended to keep ordinary citizens at a "safe" distance from the business of governing. The other shortcoming of this form of government was that it had strong affiliations with the interests of the "economic society," and thus with the interests of the wealthy classes and well-funded or well-organized interest groups who had a privileged and dominant position in it. As can be recalled, these two shortcomings were characterized earlier in this chapter as "audience democracy" and "fund democracy."

3. With all fairness to Barber, one should add that he does pay attention to questions of initiatives and national referenda. However, the problem is that he seems to view them as discrete events that would take place "some of the time" and not on a continuous basis.

4. Cohen and Rogers' notion of "associative democracy" is an exception here (Cohen and Rogers 1993).

5. Barber (1984), p.305. Again, one should take Cohen and Rogers' notion of "associative democracy" as an exception here (Cohen and Rogers 1993).

6. As was argued earlier, Habermas does address the question of sovereignty. However, he seems more interested in sublimating or desubstantializing sovereignty than actualizing it. As was discussed, contribution to decision-making (i.e., citizens' indirect participation in decision-making, and thus their indirect exercise of sovereignty) turned out to be only *potential* rather than actual. The same is the case with liberal democracy, which takes the people as the ultimate decision-makers only conceptually and figuratively by means of some convoluted arguments.

7. Gutmann and Thompson (1996), p.358 and p.8, respectively.

8. The phrase "representation gap" is borrowed from Cohen and Rogers (1993), p.246.

9. Admittedly, this is a weak argument and presupposes that deliberative democrats would agree with the proposition that the problems of epistemic differences or "moral disagreements" that they attempt to resolve or address are euphemisms for the problems posed by the power the wealthy classes wield over the political process in "fund democracy."

10. It is worth noting that theories of participatory democracy, which were by far bolder and more ambitious than theories of deliberative democracy, surged in the bold and economically prosperous decades of the 1960s and the early 1970s. By contrast, theories of deliberative democracy have flourished in the conservative and economically uncertain decade of the 1990s.

Conclusion

The history of the idea of democracy thus far has been the history of the betrayal of its moral content, and hence its ideals, especially the ideal of citizens' *direct* participation in the political process. This volume attempted to rescue the true sense of the original idea of democracy and recover its moral-political ideals from the perversions and distortions it has suffered throughout the ages. Guided by Macpherson's *Real World of Democracy* and the *Lives and Times of Democracy*, this work presented a critical examination of the hitherto existing theories and regimes of democracy.[1] As was argued, the idea of rule by the people (or the ideal of the political empowerment of the people) that prevailed in the conceptions of democracy in the pre-liberal and non-liberal societies was a problematic notion. As the idea of rule by the people, democracy was practiced successfully (albeit in a limited context) only in certain short-lived periods in ancient Athens. The success of ancient Athens in practicing democracy in those periods was attributed mainly to the presence of a general economic prosperity and the small size of its citizen body, and also in part to the absence of sharp economic inequalities among the enfranchised citizens.

Rousseau's conception of democracy, as argued, was never practiced, nor was it intended to, for his *de jure* state was a paradigmatic one. One shortcoming of Rousseau's conception was that it limited its scope to a small agrarian body politic. Although the democratic crux of Rousseau's conception, as well as its real strength, was identified in the emphasis it placed on the citizens' direct participation in decision-making, its weakness, as became apparent, lay in the fact that the citizens' participation lacked substance. The same can be said about Rousseau's emphasis on the sovereignty of the general will. Although the individual citizens in Rousseau's *de jure* state would directly vote on, or ratify in person, the laws to be legislated, the real decision-making power would rest with a charismatic and manipulative supreme leader who would intimate to citizens, through myths and religious deception, what their general will would be. In contrast to Rousseau's conception of democracy, the Leninist idea of "proletarian democracy" was put into practice on a grand scale, as it was intended—and ended up producing undemocratic results. The failure of the idea of proletarian democracy was attributed, in part, to some of the theoretical shortcomings inherent in the idea itself.

Painting with a broad brush, all of the pre-liberal and non-liberal conceptions of democracy either had limited scope and applicability, in particular those of ancient Athens and Rousseau, or were flawed theoretically as was the case with ancient Athens' weakness on the question of equality.[2] The theoretical flaws in the case of Rousseau were outlined just above. As to Bolshevism, the theoretical flaws of the idea of the proletarian democracy can be summed up as: the abandonment of the idea of direct Soviet democracy; the absence of checks and balances, and democratic procedures needed for settling political disputes; and the willingness to trample upon individual liberties. As was suggested in the concluding section of Part I, these shortcomings and flaws were directly related, for the most part, to the historical-developmental limitations that served as contexts for the pre-liberal and non-liberal conceptions of democracy. These limitations were of two types: underdeveloped modes of economic production, and hence the circumstances of scarcity, on the one hand, and the underdeveloped political-legal institutions and cultural attitudes that accompanied these circumstances, on the other hand.

In addition to examining the history of the idea of democracy in the pre-liberal and non-liberal societies, the present volume also studied the history of the idea in liberal and liberal-democratic societies. If the perversion of the original idea in the pre-liberal and non-liberal era could be blamed, for the most part, on the circumstances of scarcity and their accompanying social-political and cultural shortcomings, the blame for the perversion of the original idea in the liberal and liberal-democratic societies should be placed largely on the shrewd efforts that consciously sought to use democracy as a stabilizing strategy. As far as the propertied classes of the liberal society were concerned, social upheavals had to be stopped, and revolutions pre-empted or stalled. The liberal society had to be made safe for free-markets. The liberalization of democracy meant that the idea had to be rehabilitated into a "good thing" and be admitted into the realm of the respectable. This required that democracy be beaten into a shape that would make it suitable for the purpose of presenting it as a category of the idea of freedom that was worshiped zealously by the liberal state and its free-market ideologues. It was in this process that democracy lost its moral content and was contorted into a "free" and anormative "method" of decision-making. This undertaking in the period of the late-nineteenth to early-twentieth centuries was largely guided and carried out by the class-conscious and well-organized elements of the propertied classes and their allies, primarily the intellectual elites of the middle-classes, who were moved into action by their desire to protect their interests and their fears that they might lose their privileged status to the "undesirable elements," the lower classes, who were waging class wars against the liberal state.

Whereas the liberal society in its earlier stages made no pretense of being democratic, in that it did not embrace the principle of universal suffrage, the liberal-democratic society portrayed itself as democracy *par excellence*. As was discussed, the underlying problem with the "liberal-democratic conception of democracy" is that it relies exclusively on representation, and thus contorts the idea of democracy into the idea of rule by a freely and popularly elected representative government. The main difficulty with this conception is that it sidelines citizens politically, and transfers the legislative and public policy decision-making powers into the hands of their representatives—indeed a class of political elites and "natural aristocrats" who gain this status by winning elections. The idea of representative government in the liberal and liberal-democratic states was premised on keeping citizens away from the business of governing. Moreover, it was argued that representative governments in both states were instrumentalized by the wealthy classes and well-financed or well-organized special interests groups who sought to use the state for their socio-economic and political purposes. Finally, the liberal-democratic conception is democratic mainly in the sense that it arranges the transfer of the decision- and policy-making powers from the people to the officials via the method of "free" and "competitive" elections. Not the question of the political empowerment of the people, but rather the idea of their "democratic" *disempowerment* that serves as the linchpin of the "liberal-democratic conception of democracy."

As a way of redressing these problems, theories of participatory democracy and deliberative democracy set out to carve out a space for citizens' participation in the liberal-democratic conception of democracy. One focus of these theories, especially the theories of participatory democracy, was to undermine what they regarded as the anti-democratic biases of the liberal-democratic conceptions of individual rights. The low appeal and the eventual decline of the idea of participatory democracy, it was argued, should be attributed primarily to its underestimation of the power of liberal values in the liberal-democratic society, as well as to its attempts to undermine the preponderance of the "liberal" pole of the theory as a way of breathing substance into its "democratic" pole. Moreover, although theories of deliberative democracy managed to sidestep this pitfall, they suffered from a host of other difficulties that also haunted the theories of participatory democracy. Leaving aside the problems associated with model-making and the actual practice of democracy in deliberative ways, theories of deliberative democracy failed, as their precursors did, to restore to democracy its main and most pressing ideal, viz., the ideal that democracy is primarily about demos being the sovereign in the state and exercising this sovereignty *directly*—that *the people themselves should be the actual decision-makers*—or that they should have the power to ratify or reject, "in person," the decisions made on their behalf by their appointees.

Building on what this volume presented, *Democracy as the Political Empowerment of the Citizen: Direct-Deliberative e-Democracy* attempts to offer a new political-philosophical framework that will make it possible to formulate a new theory of democracy that would restore to the concept the full range of its original ideals. As the title indicates, the companion volume reformulates democracy as the question of the political empowerment of the individual citizens who comprise "the people." Moreover, as the subtitle suggests, the new formulation resorts to the new idea of "*e*-democracy" in order to find an alternative approach to the question of making the ancient ideals of the citizens' direct, deliberative, and equal participation relevant to the realties of the modern nation-state.

An important part of the project of the companion volume is the contention that present-day society is much better suited for realizing the original idea of democracy than ancient Athens itself was in its most democratic moments. Present-day society has at its disposal the material, technological, social-political, and cultural-educational infrastructures and institutions needed to confront the factors that contributed to the perversion of the original idea in the earlier periods. These infrastructures and institutions, and a popular democratic will that they can help foster, put our generation in the unique position of being the first to set for itself the task of establishing the first truly democratic society in history.

Notes

1. As was seen, in retrieving the full scope of the moral substance of democracy, the book went beyond the attempt to merely restore fullness to the egalitarian component of democracy (which was the main focus of Macpherson's retrieval project), and also attempted to reclaim for democracy its ideals of the citizens' participation in public deliberations and their direct contributions to the political decision-making process.

2. Ancient Athens' weakness on the question of equality had two dimensions. First, its notion of equality was formal-legal in character, and thus lacked the "substance" one finds in Rousseau's conception. (Well-to-do dominated the Council of 500, and middle-class individuals the Assembly and the juries, and the poor either did not participate or were abused when they did—see notes 2 and 3 in Chapter 2. One should add that this weakness did not pose a serious problem for Athenian democracy in its glory moments, when the franchise paid citizens for attending the assembly, and there existed a general economic prosperity.) Second, Athenian democracy treated non-Greeks as having lesser status than Greeks, thus excluding them, alongside women and slaves, from the franchise.

Selected Bibliography

Albert, Michel. *Capitalism vs. Capitalism: How America's Obsession with Individual Achievement and Short-term Profit Has Led It to the Brink of Collapse*. New York: Four Wall Eight Windows, 1993.

Allen, Anita L., and Milton C. Regan, Jr., eds. *Debating Democracy's Discontent: Essay on American Politics, Law, and Public Philosophy*. Oxford: Oxford University Press, 1998.

Ankersmit, F. R. *Political Representation*. Stanford: Stanford University Press, 2002.

Appleby, Joyce, Lynn Hunt, and Margaret Jacobs. *Telling the Truth about History*. New York: W. W. Norton and Company, 1994.

Archer, Robin. *Economic Democracy: The Politics of Feasible Socialism*. Oxford: Clarendon Press, 1995.

Archer, Robin, Diemut Bubeck, Hanjo Glock, Lesley Jacobs, Seth Moglen, Adam Steinhouse, and Daniel Weinstock, eds. *Out of Apathy: Voices of the New Left Thirty Years On*. London: Verso, 1989.

Aristotle. *The Politics of Aristotle*. Translated and with an introduction, notes and appendixes by Ernest Barker. Oxford: Oxford University Press, 1958.

Arthur, John, ed. *Democracy: Theory and Practice*. Belmont, CA: Wadsworth Publishing Company, 1992.

Avon, Dan, and Anver Shalit, eds. *Liberalism and Its Practice*. London: Routledge, 1999.

Bachrach, Peter, and Aryeh Botwinick, eds. *Power and Empowerment: A Radical Theory of Participatory Democracy*. Philadelphia: Temple University Press, 1992.

Baker, C. Edwin. *Media, Markets, and Democracy*. Cambridge: Cambridge University Press, 2002.

Baker, Earnest. *The Political Thought of Plato and Aristotle*. New York: Dover Publications, Inc., 1959.

Ball, Terence, James Farr, and Russell L. Hanson, eds. *Political Innovation and Conceptual Change*. Cambridge: Cambridge University Press, 1989.

Baran, Paul, and Paul Sweezy. *Monopoly Capital: An Essay on the American Economic and Social Order*. New York: Monthly Press Review, 1968.

Barber, Benjamin R. *Strong Democracy.* Berkeley: University of California Press, 1984.

———. *A Passion for Democracy.* Princeton: Princeton University Press, 1998.

———. "Three Scenarios for the Future of Technology and Strong Democracy." *Political Science Quarterly.* vol.113, no.4 (1998-99): 573-89.

Beard, Charles. *An Economic Interpretation of the Constitution of the United States.* New York: Macmillan Company, 1935.

Beaud, Michel. *A History of Capitalism 1500-2000.* Translated and edited by Tom Dickman and Anny Lefebvre. New York: Monthly Review Press, 2001.

Becker, Barbara, and Josef Wehner. "Electronic Networks and Civil Society: Reflections on Structural Changes in the Public Sphere." In *Culture, Technology, Communication: Toward an Intercultural Global Village.* Edited by Charles Ess and Fay Sudweeks. Albany, NY: State University of New York Press, 2001.

Behrouzi, Majid. *Democracy as the Political Empowerment of the Citizen: Direct-Deliberative e-Democracy.* Lanham, MD: Lexington Books, 2005.

Beitzinger, A. J. *A History of American Political Thought.* New York: Dodd, Mead and Company, 1972.

Bender, Frederic L. *The Betrayal of Marx.* New York: Harper & Row Publishers, 1975.

Benello, C. George. *From the Ground Up: Essays on Grassroots and Workplace Democracy.* Boston: South End Press, 1992.

Berlin, Isaiah. *Four Essays on Liberty.* Oxford: Oxford University Press, 1969.

Berry, Christopher J. *The Idea of a Democratic Community.* New York: St. Martin's Press, 1989.

Birch, Anthony H. *Representation.* New York: Praeger, 1971.

Blaug, Richard. "New Theories of Discursive Democracy: A User's Guide." *Philosophy and Social Criticism,* vol.22, no.1 (1996): 49-80.

———. *Democracy, Real and Ideal: Discourse Ethics and Radical Politics.* Albany, NY: State University of New York Press, 1999.

Boardman, John, Jasper Griffin, and Oswyn Murray, eds. *The Oxford History of the Classical World.* Oxford: Oxford University Press, 1986.

Bloom, Allan. "Rousseau's Critique of Liberal Constitutionalism." In *The Legacy of Rousseau.* Edited by Clifford Orwin, and Nathan Tarcov. Chicago: The University of Chicago Press, 1997.

Bohman, James. "Survey Article: The Coming of Age of Deliberative Democracy." *The Journal of Political Philosophy,* vol.6, no.4 (1998): 400-425.

Bonner, Robert J. *Aspects of Athenian Democracy.* New York: Russell & Russell, 1967.

Bottomore, Tom, Laurence Harris, V. G. Kiernan, and Ralph Miliband, eds. *A Dictionary of Marxist Thought.* Cambridge, MA: Harvard University Press, 1983.

Bowles, Samuel, and Herbert Gintis. *Democracy and Capitalism: Property, Community, and the Contradictions of Modern Social Thought.* New York: Basic Books, 1986.

Brink, Bert Van Den. *The Tragedy of Liberalism: An Alternative Defense of a Political Tradition.* Albany, NY: State University of New York Press, 2000.

Broder, David S. *Democracy Derailed: Initiative Campaigns and the Power of Money.* New York: Harcourt Inc., 2000.

Brown, A. "Reconstructing the Soviet Political System." In *Chronicle of a Revolution: A Western-Soviet Inquiry Into Perestroika.* Edited by A. Brumberg. New York: Pantheon, 1990.

Buchanan, James. "Social Choice, Democracy, and Free Markets." *Journal of Political Economy,* April 1954, 62(2), pp.114-23.

Buchanan, James M., and Gordon Tullock. *The Calculus of Consent: Logical Foundations of Constitutional Democracy.* Ann Arbor: University of Michigan Press, 1965.

Budge, Ian. *The New Challenge of Direct Democracy.* Cambridge: Polity Press, 1996.

Burke, Edmund. *The Writing and Speeches of the Right Honorable Edmund Burke.* Beaconsfield edition in 12 vols. Boston: Little, Brown and Company, 1901.

Burnheim, John. *Is Democracy Possible?* Berkeley: University of California Press, 1985.

Calhoun, Craig. "The Public Good as a Social and Cultural Project." In *Private Action and the Public Good.* Edited by Walter W. Powell and Elizabeth S. Clemens. New Haven: Yale University Press, 1998.

Carens, Joseph H., ed. *Democracy and Possessive Individualism: The Intellectual Legacy of C. B. Macpherson.* Albany, NY: State University of New York Press, 1993.

Chapman, John W., and I. Shapiro, eds. *Democratic Community.* New York: New York University Press, 1993.

Chomsky, Noam. *Profit Over People.* New York: Seven Stories Press, 1999.

Cladis, Mark S. *Public Visions, Private Lives: Rousseau, Religion, and 21st-Century Democracy.* Oxford: Oxford University Press, 2003.

Claster, Jill N., ed. *Athenian Democracy: Triumph or Travesty?* New York: Holt, Rinehart and Winston, 1967.

Cohen, Carl, ed. *Communism, Fascism, and Democracy: The Theoretical Foundations.* New York: McGraw-Hill, 1997.

Cohen, Joshua. "Democracy and Liberty." In *Deliberative Democracy.* Edited by John Elster. Cambridge: Cambridge University Press, 1998.

———. "An Epistemic Conception of Democracy." *Ethics,* vol.97 (October 1986): 26-38.

———. "Deliberation and Democratic Legitimacy." In *Deliberative Democracy: Essays on Reason and Politics,* edited by James Bohman and William Rehg. Cambridge, MA: MIT Press, 1997a.

——. "Procedure and Substance in Deliberative Democracy." In *Deliberative Democracy: Essays on Reason and Politics*, edited by James Bohman and William Rehg. Cambridge, MA: MIT Press, 1997b.

Cohen, Joshua, and Joel Rogers. "Associative Democracy." In *Market Socialism: The Current Debate*, edited by Pranab K. Bardhan and John E. Roemer. Oxford: Oxford University Press, 1993.

——. *Associations and Democracy: The Utopian Project*, vol.1, edited by Erik Olin Wright. London: Verso, 1995.

——. "My Utopia or Yours?" In *Equal Shares: Making Market Socialism Work the Real Utopias Project*, vol.II, by John E. Roemer, edited and introduced by Erik Olin Wright. London: Verso, 1996.

Colletti, L. *From Rousseau to Lenin*. London: New Left Books, 1972.

Connolly, William E. *The Terms of Political Discourse*. Second ed. Princeton, NJ: Princeton University Press, 1983.

Constant, Benjamin. *The Political Writings of Benjamin Constant*. Edited by Biancamaria Fontona. Cambridge: Cambridge University Press, 1988.

Copp, David, Jean Hampton, and John Roemer, eds. *The Ideal of Democracy*. Cambridge: Cambridge University Press, 1993.

Corlett, J. Angelo. "Marx and Rights." *Dialogue* (Canada), XXXIII, (1994).

Cowling, Mark, and Paul Reynolds, eds. *Marxism, the Millennium and Beyond*. London: Palgrave Publishers Ltd., 2000.

Craig, Edward, gen. ed. *Routledge Encyclopedia of Philosophy*. London: Routledge, 1998.

Craig, Stephen C., ed. *Broken Contract? Changing Relations Between Americans and Their Government*. Boulder, CO: Westview Press, 1996.

Cunningham, Frank. *Democratic Theory and Socialism*. Cambridge: Cambridge University Press, 1987.

——. *The Real World of Democracy Revisited*. Atlantic Highlands, NJ: Humanities Press, 1994.

——. *Theories of Democracy: A Critical Introduction*. London: Routledge, 2002.

Dagger, Richard. *Civic Virtues: Rights, Citizenship, and Republican Liberalism*. New York: Oxford University Press, 1997.

Dahl, Robert A. *A Preface to Economic Theory*. Chicago: The University of Chicago Press, 1956.

——. *Democracy and Its Critics*. New Haven, CT: Yale University Press, 1989.

——. *On Democracy*. New Haven, CT: Yale University Press, 1998.

Dahl, Robert, Ian Shapiro, Jose Antonio Cheibub. *The Democracy Source Book*. Cambridge, MA: MIT Press, 2003a.

Dalton, Russell J., Wilhelm Burklin, Andrew Drummond. "Public Opinion and Direct Democracy." *Journal of Democracy*, vol.12, no.4, (October 2001).

Dante, Germino. *Machiavelli to Marx: Modern Western Political Thought*. Chicago: The University of Chicago Press, 1972.

Delanty, Gerard. *Community*. London: Routledge, 2003.

DeMarco, Joseph P., and Samuel A. Richmond. "The Mutuality of Liberty, Equality, and Fraternity." *Journal of Social Philosophy* 17 (Fall 1986): 7-12.

Denning, S. Lance. *The Practice of Workplace Participation.* Westport, CT: Quorum Books, 1998.

Deutscher, Isaac. *The Profit Unarmed, Trotsky: 1921-1929.* London: Oxford University Press, 1959.

Dew, John R. *Empowerment and Democracy in the Workplace.* Westport, CT: Quorum Books, 1997.

Dewey, John. *Philosophy of Education.* Totowa, NJ: Adams and Company, 1975 [1958].

——. *Liberalism and Social Action.* New York: Perigee Books, 1980.

——. *The Public and Its Problems.* Chicago: Swallow Press, 1954 [1927].

——. "The Ethics of Democracy." In *Jon Dewey: The Political Writings*, edited and introduced by Debra Morris and Ian Shapiro. Indianapolis, IN: Hackett Publishing Company, 1993 [1888].

Donnelly, David, Janice Fine, and Ellen S. Miller. *Are Elections for Sale?* Boston: Beacon Press, 2001.

Downs, Anthony. *An Economic Theory of Democracy.* New York: Harper and Row, 1957.

Draper, Hal. *The "Dictatorship of the Proletariat" from Marx to Lenin.* New York: Monthly Review Press, 1987.

Drew, Elizabeth. *The Corruption of American Politics: What Went Wrong and Why.* New York: The Overlook Press, 2000.

——. *Politics and Money.* New York: Macmillan Publishing Company, 1983.

Dryzek, John S. *Deliberative Democracy and Beyond: Liberals, Critics, Contestations.* Oxford: Oxford University Press, 2000.

——. "Legitimacy and Economy in Deliberative Democracy." *Political Theory*, vol.29, no.5 (October 2001): 651-69.

DuBoff, Richard B. *Accumulation and Power: An Economic History of the United States.* Armonk, NY: M. E. Sharpe Inc., 1989.

Dummett, Michael. *Voting Procedures.* Oxford: Clarendon Press, 1984.

Duncan, Christopher M. *The Anti-Federalists and Early American Political Thought.* DeKalb, IL: Northern Illinois University Press, 1995.

Dunn, John, ed. *Democracy the Unfinished Journey: 508 BC to AD 1993.* Oxford: Oxford University Press, 1992.

Dworkin, Ronald. *A Matter of Principle.* Cambridge, MA: Harvard University Press, 1985.

——. *Sovereign Virtue.* Cambridge, MA: Harvard University Press, 2000.

Dye, Thomas R. *Who's Running America? The Clinton Years.* Sixth edition. Englewood Cliffs, NJ: Prentice Hall, 1995.

Edwards, Paul (chief editor). *The Encyclopedia of Philosophy.* New York: Macmillan Publishing Company, 1967.

Eisenberg, Avigail I. *Reconstructing Political Liberalism.* Albany, NY: State University of New York Press, 1995.

Ellul, Jacques. *The Technological Society* (1954). Translated by John Wilkinson. New York: Vintage Books, 1964 (1954).

——. "Technology and Democracy." In *Democracy in a Technological Society*, edited by Langdon Winner. Boston: Kluwer Academic Publishers, 1992.

Elster, Jon, and Aanurd Hylland, eds. *Foundations of Social Choice Theory.* Cambridge: Cambridge University Press, 1986.

Elster, Jon, ed. *Deliberative Democracy.* Cambridge: Cambridge University Press, 1998.

——. "Possibility of Rational Politics." In *Political Theory Today*, edited by David Held. Stanford: Stanford University Press, 1991.

Estlund, David M. "Who's Afraid of Deliberative Democracy? On the Strategic/Deliberative Dichotomy in Recent Constitutional Jurisprudence." *Texas Law Review,* vol.71 (1993): 1437-77.

Etzioni, Amitai. *New Communitarian Thinking: Persons, Virtues, Institutions, and Communities.* Charlottesville: University Press of Virginia, 1995.

Farber, Samuel. *Before Stalinism: The Rise and Fall of Soviet Democracy.* London: Verso, 1990.

Femia, Joseph V. *Marxism and Democracy.* Oxford: Clarendon Press, 1993.

——. "Complexity and Deliberative Democracy." *Inquiry* 39 (1996): 359-97.

Fishkin, James S. *Democracy and Deliberation.* New Haven: Yale University Press, 1991.

Flower, Robert B. *The Dance with Community.* Lawrence: University Press of Kansas, 1991.

Freeman, Samuel. "Deliberative Democracy: A Sympathetic Comment." *Philosophy and Public Affairs* 29, no.4 (Fall 2000): 371-418.

Gallie, W. B. *Philosophy and the Historical Understanding.* Second ed. New York: Schocken Books, 1968.

Germino, Dante. *Machiavelli to Marx: Modern Western Political Thought.* Chicago: The University of Chicago Press, 1979.

Golding, Sue. *Gramsci's Democratic Theory: Contributions to a Post-Liberal Democracy.* Toronto: University of Toronto Press, 1992.

Gould, Carol C. *Rethinking Democracy.* Cambridge: Cambridge University Press, 1988.

Grady, Robert C. *Restoring Real Representation.* Urbana, IL: University of Illinois Press, 1993.

Gramsci, A. *Selections from Prison Notebooks.* New York, International Publishers, 1971.

Gray, John. *Liberalism.* Minneapolis: University of Minnesota Press, 1995.

Green, Judith M. *Deep Democracy: Community, Diversity, and Transformation.* Lanham, MD: Rowman & Littlefield Publishers, Inc., 1999.

Green, Leslie. "Dictators and Democracies." *Analysis*, vol.43, no.1 (January 1983): 58-59.

Green, Mark. *Selling Out: How Big Corporate Money Buys Elections, Rams Through Legislation, and Betrays Our Democracy.* New York: Regan Books, 2004.

Greenberg, Edward S. *Workplace Democracy: The Political Effects of Participation.* Ithaca, NY: Cornell University Press, 1986.

Greenberg, Stanley B. *The Two Americas: Our Current Political Deadlock and How to Break it.* New York: Thomas Dunne Books, 2004.

Gundersen, Adolf G. *The Socratic Citizens: A Theory of Deliberative Democracy.* Lanham, MD: Lexington Books, 2000.

Gutmann, Amy, and Dennis Thompson. *Democracy and Disagreement.* Cambridge, MA: Harvard University Press, 1996.

Habermas, Jurgen. *Legitimation Crisis.* Translated by Thomas McCarthy. Boston: Beacon Press, 1975.

———. "Three Normative Models of Democracy." *Constellations,* vol.1, no.1 (1994): 1-10.

———. "Human Rights and Popular Sovereignty: The Liberal and Republican Versions." In *Human Rights Law*, edited by Philip Alston. New York: New York University Press, 1996.

———. "Popular Sovereignty and Procedure." In *Deliberative Democracy: Essays on Reason and Politics*, edited by James Bohman and William Rehg. Cambridge, MA: MIT Press, 1997.

———. *Between Facts and Norms.* Cambridge, MA: MIT Press, 1998.

Hamilton, Alexander, James Madison, and John Jay. *The Federalist.* Edited by Benjamin Fletcher Wright. Cambridge, MA: The Belknap Press of Harvard University Press, 1961.

Hanson, Russell L. "Democracy." In *Political Innovation and Conceptual Change*, edited by Terence Ball, James Farr, and Russell L. Hanson. Cambridge: Cambridge University Press, 1989.

Haumptmann, Emily. *Putting Choice Before Democracy.* Albany, NY: State University of New York Press, 1996.

Held, David. *Models of Democracy.* Palo Alto, CA: Stanford University Press, 1989.

Hertz, Noreena. *The Silent Takeover: Global Capitalism and the Death of Democracy.* London: Heinemann, 2001.

———. *Prospects for Democracy: North South East West.* Stanford: Stanford University Press, 1993.

———. *Democracy and the Global Order.* Cambridge: Polity Press, 1995.

Hibbing, John R., and Elizabeth Theiss-Morse. *What Is It About Government That Americans Dislike?* London: Cambridge University Press, 2001.

Hightower, Jim. *Thieves in High Places: They've Stolen Our Country and It's Time to Take It Back.* New York: Viking, 2003.

Hirst, Paul. *Representative Democracy and Its Limits.* Cambridge: Polity Press, 1990.

Hobbes, Thomas. *Leviathan* (1651). Edited by Richard Tuck. Cambridge: Cambridge University Press, 1996.

Hoffman, Ronald, and Peter J. Albert, eds. In *The Transforming Hand of Revolution: Reconsidering the American Revolution as a Social Movement.* Charlottesville: The University Press of Virginia, 1996.

Howard, Michael W. *Self-Management and the Crisis of Socialism: The Rose in the Fist of the Present*. Lanham, MD: Rowman and Littlefield Publishers, Inc., 2000.

——, ed. *Socialism*. Amherst, MA: Humanity Books, 2001.

Howard, Rhoda E. *Human Rights and Search for Community*. Oxford: Westview Press, 1995.

Huffington, Arianna. *How to Overthrow the Government*. New York: Harper Collins Publishers, 2000.

——. *Pigs at the Trough: How Corporate Greed and Political Corruption Are Undermining America*. New York: Crown Publishers, 2003.

Isaac, Jeffrey C. *Democracy in Dark Times*. Ithaca, NY: Cornell University Press, 1988.

Jacobs, Lesley A. *An Introduction to Modern Political Philosophy*. Upper Saddle River, NJ: Prentice Hall, 1997.

——. "Market Socialism and Non-Utopian Marxist Theory." *Philosophy of Social Sciences,* vol.29, no.4 (December 1999): 527-39.

Jaggar, Allison M. "Multicultural Democracy." *The Journal of Political Philosophy,* vol.7, no.3 (1999): 308-29.

Jones, A. H. M. *Athenian Democracy*. Oxford: Basil Blackwell, 1964.

Judis, John B. *The Paradox of American Democracy*. New York: Routledge, 2000.

Kant, Immanuel. *Foundations of the Metaphysics of Morals* (1785). Translated and Introduced by Lewis White Beck. New York: Macmillan Publishing Company, 1959.

Kautz, Steven. *Liberalism and Community*. Ithaca, NY: Cornell University Press, 1995.

Kegley, Jacquelyn A. K. *Genuine Individuals and Genuine Communities*. Nashville: Vanderbilt University Press, 1997.

Kenez, Peter. *The Birth of the Propaganda State*. Cambridge: Cambridge University Press, 1985.

Knei-Paz, Baruch. "Was George Orwell Wrong?" *Dissent* 42 (2) (Spring 1995): 266-69.

Knight, Jack, and James Johnson. "Aggregation and Deliberation: On the Possibility of Democratic Legitimacy." *Political Theory,* vol.22, no.2 (May 1994): 277-96.

Kochler, Hans. *The Crisis of Representative Democracy*. Frankfurt: Verlag Peter Lang, 1987.

Krimerman, Len, and Frank Lindenfeld, eds. *When Workers Decide: Workplace Democracy Takes Root in North America*. Philadelphia: New Society Publishers, 1992.

Kulikoff, Allan. *The Agrarian Origins of American Capitalism*. Charlottesville: University Press of Virginia, 1992.

Kymlicka, Will. *Liberalism, Community and Culture*. Oxford: Clarendon Press, 1981.

——. *Contemporary Political Philosophy.* New York: Oxford University Press, 1990.

——. *Finding Our Way: Rethinking Ethnocultural Relations in Canada.* Oxford: Oxford University Press, 1998.

——. "Liberal Egalitarianism and Civic Republicanism: Friends or Enemies?" In *Debating Democracy's Discontent: Essays on American Politics, Law, and Public Philosophy,* edited by Anita L. Allen and Milton C. Regan. Oxford: Oxford University Press, 1998.

——. "American Multiculturalism and the 'Nation Within.'" In *Political Theory and the Rights of Indigenous Peoples,* edited by Ivison Duncan, Paul Patton, and Will Sanders. Cambridge: Cambridge University Press, 2000.

——. *Contemporary Political Philosophy.* Oxford: Oxford University Press, 2002.

Lane, David. *Leninism: A Sociological Interpretation.* Cambridge: Cambridge University Press, 1981.

Lenin, V. I. *Selected Works in Two Volumes.* Moscow: Foreign Languages Publishing House, 1952.

——. *Collected Works.* Moscow: Progress Publishers, 1976.

Levine, Andrew. *The Politics of Autonomy: A Kantian Reading of Rousseau's Social Contract.* Amherst, MA: University of Massachusetts Press, 1976.

——. *Liberal Democracy: A Critique of Its Theory.* New York: Columbia University Press, 1981.

——. *Arguing for Socialism.* Boston, London: Routledge & Kegan Paul, 1984.

——. *The General Will: Rousseau, Marx, Communism.* Cambridge: Cambridge University Press, 1993.

——. "Democratic Corporatism and/versus Socialism." In *Associations and Democracy: The Utopian Project,* vol.1. Edited by Joshua Cohen and Joel Rogers. London: Verso, 1995.

——. *Engaging Political Philosophy: From Hobbes to Rawls.* Oxford: Blackwell Publishers, 2002.

Lichterman,, Paul. *The Search for Community: American Activists Reinventing Commitment.* Cambridge: Cambridge University Press, 1996.

Lindsay, Peter. *Creative Individualism.* Albany, NY: State University of New York Press, 1996.

Lipset, Seymour Martin (chief editor). *The Encyclopedia of Democracy.* In 4 vols. Washington, D.C.: Congressional Quarterly Inc., 1995.

List, Christian, and Robert E. Goodin. "Epistemic Democracy: Generalizing the Condorcet Jury Theorem." *The Journal of Political Philosophy,* vol.9, no.3 (2001): 277-306.

Lopston, Peter. *Theories of Human Nature.* Peterborough, ON: Broadview Press, 1995.

Lovell, David W. *From Marx to Lenin: An Evaluation of Marx's Responsibility for Soviet Authoritarianism.* Cambridge: Cambridge University Press, 1984.

Lummis, C. Douglas. *Radical Democracy.* Ithaca, NY: Cornell University Press, 1996.

Luxemburg, Rosa. *Rosa Luxemburg Speaks*. Edited by Mary-Alice Waters. New York: Pathfinder Press, Inc., 1970.

Macedo, Stephen, ed. *Deliberative Politics: Essays on Democracy and Disagreement*. Oxford: Oxford University Press, 1999.

Machiavelli, Niccolo. *The Prince*. Edited with an introduction by Peter Bondanella, Translated by Peter Bondanella and Mark Musa. Oxford: Oxford University Press, 1984.

Macpherson, C. B. *The Real World of Democracy*. Toronto: House of Anansi Press, 1987 [1965].

———. *Democratic Theory: Essays in Retrieval*. Oxford: Clarendon Press, 1973.

———. *The Life and Times of Liberal Democracy*. Oxford: Oxford University Press, 1977.

Manin, Bernard. *The Principles of Representative Government*. Cambridge: Cambridge University Press, 1997.

Mansbridge, Jane. *Beyond Adversary Democracy*. New York: Basic Books, 1980.

———. "Reconstructing Democracy." In *Revisioning the Political: Feminist Reconstructions of Traditional Concepts in Western Political Theory*, edited by Nancy Hirschmann and D. Stefano. Boulder, CO: Westview Press, 1996.

———. "On the Contested Nature of the Public Good." In *Private Action and the Public Good*, edited by Walter W. Powell and Elisabeth S. Clemens. New Haven: Yale University Press, 1998.

Marcuse, Herbert. *Reason and Revolution: Hegel and the Rise of Social Theory* (1954). 2nd ed. with supplementary chapter. Atlantic Highlands, NJ: Humanities Press, Inc., 1983.

———. *Eros and Civilization: A Philosophical Inquiry into Freud* (1955). Boston: Beacon Press, 1966.

———. *One Dimensional Man: Studies in the Ideology of Advanced Industrial Society*. Boston: Beacon Press, 1964.

Marx, Karl. *Karl Marx: Early Writings*. Translated and edited by T. B. Bottomore. New York: McGraw Hill, 1963.

———. *Critique of Hegel's "Philosophy of Right."* Edited by Joseph O'Malley. Cambridge: Cambridge University Press, 1970 [1859].

———. *The Letters of Karl Marx*. Edited by Saul K. Padover. Englewood Cliffs, NJ: Prentice Hall, 1979.

Marx, Karl, and Fredrick Engels. *Selected Works in Three Volumes*. Moscow: Progress Publishers, 1977.

———. *The German Ideology*. Moscow: Progress Publishers, 1976.

———. *Collected Works*. New York: International Publishers, 1975.

McClure, Kristie. "On the Subject of Rights: Pluralism, Plurality and Political Identity." In *Dimensions of Radical Democracy: Pluralism, Citizenship, Community*, edited by Chantal Mouffe. London: Verso, 1992.

McLean, Iain. *Dealing in Votes: Interactions Between Politicians and Voters in Britain and the USA*. Oxford: Martin Robertson, 1982.

———. *Public Choice: An Introduction*. Oxford: Basil Blackwell, 1987.

———. *Democracy and New Technology*. Cambridge: Polity Press, 1989.

———. "Rational Choice and Politics." *Political Studies*, vol.xxxix (1991a): 496-512.

———. "Forms of Representation and Systems of Voting." In *Political Theory Today*, edited by David Held. Stanford: Stanford University Press,1991b.

McLean, Iain, and Arnold B. Urken., eds. *Classics of Social Choice*. Ann Arbor: University of Michigan Press, 1995.

Melman, Seymour. *After Capitalism: From Managerialism to Workplace Democracy*. New York: Alfred A. Knopf, 2001.

Melzer, Arthur M. *The Natural Goodness of Man: On the System of Rousseau's Thought*. Chicago: The University of Chicago Press, 1990.

Miliband, Ralph. *The State in Capitalist Society*. London: Quartet Books, 1980.

Mill, John Stuart. *Considerations on Representative Government* (1861). Edited by Currin V. Shields. New York: The Bobbs-Merrill Company, Inc., 1958.

Miller, David. *Market, State, and Community: Theoretical Foundations of Market Socialism*. Oxford: Clarendon Press, 1989.

———. *Citizenship and National Identity*. Cambridge: Polity Press, 2000.

Miller, James. *"Democracy Is in the Streets": From Port Huron to the Siege of Chicago*. New York: Simon and Schuster, 1987.

Miller, Joshua. *The Rise and Fall of Democracy in Early America, 1630-1789*. University Park: The Pennsylvania State University Press, 1991.

Montesquieu, Charles-Louis de Secondat. *The Spirit of the Laws*. Translated and edited by Anne M. Cohler, Basia Carolyn Miller, and Harold Samuel Stone. Cambridge: Cambridge University Press, 1989.

Morera, Esteve . *Gramsci's Historicism*. London: Routledge, 1990.

———. "Gramsci and Democracy." *Canadian Journal of Political Science*, XXIII:1 (March 1990), 23-37.

Morrison, John. "The Case Against Constitutional Reform?" *Journal of Law and Society*, vol.25, no.4 (December 1998): 510-35.

Mouffe, Chantal, and Ernesto Laclua. *Hegemony and Socialist Strategy: Toward a Radical Democratic Politics*. London: Verso, 1985.

Mouffe, Chantal, ed. *Dimensions of Radical Democracy: Pluralism, Citizenship, Community*. London: Verso, 1992.

———. *The Return of the Political*. London: Verso, 1993.

Mukherjee, Rabin. *Democracy: A Failure, Shefocracy: The Solution for Human Welfare*. New York: University Press of America, 2000.

Nader, Ralph. *Crashing the Party: Taking on the Corporate Government in an Age of Surrender*. New York: St. Martin's Press, 2002.

Nedelsky, Jennifer. *Private Property and the Limits of American Constitutionalism*. Chicago: The University of Chicago Press, 1990.

Neidleman, Jason A. *The General Will Is Citizenship: Inquiries into French Political Thought*. New York: Rowman and Littlefield Publishing Co., 2001.

Newman, Bruce I. *The Mass Marketing of Politics: Democracy in an Age of Manufactured Images*. Thousand Oaks, CA: Sage Publications, 1999.

Nino, Carlos Santiago. *The Constitution of Deliberative Democracy.* New Haven: New York University Press, 1996.

Nun, Jose. *Democracy: Government of the People or Government of the Politicians?* Lanham, MD: Rowman & Littlefield Publishers, 2003.

Ollman, Bertell, ed. *Market Socialism: The Debate Among the Socialists.* New York: Routledge, 1998.

O'Shaughnessy, Nicholas, J. *The Phenomenon of Political Marketing.* London: MacMillan Press, 1990.

Pagano, Ugo, and Robert Rowthorn, eds. *Democracy and Efficiency in the Economic Enterprise.* London: Routledge, 1996.

Palast, Greg. *The Best Democracy Money Can Buy.* London: Pluto Press, 2002.

Palmer, R. R. "Notes on the Use of the Word 'Democracy' 1789-1799." *Political Science Quarterly,* vol.68 (1953): 203-26.

Pangle, Thomas L. *Montesquieu's Philosophy of Liberalism: A Commentary on The Spirit of the Laws.* Chicago: The University of Chicago Press, 1973.

Parenti, Michael. *Democracy for the Few.* New York: St. Martin's Press, 1996.

———. *The Assassination of Julius Caesar: A People's History of Ancient Rome.* New York, London: The New Press, 2003.

Pateman, Carole. *Participation and Democratic Theory.* Cambridge: Cambridge University Press, 1970.

———. "Social Choice or Democracy? A Comment on Coleman and Ferejohn." *Ethics,* vol.97 (October 1986): 39-46.

Peffer, R. G. *Marxism, Morality, and Social Justice.* Princeton: Princeton University Press, 1990.

Pettit, Philip. "Reworking Sandel's Republicanism." In *Debating Democracy's Discontent: Essays on American Politics, Law, and Public Philosophy,* edited by Anita L. Allen and Milton C. Regan. Oxford: Oxford University Press, 1998.

———. "The Virtual Reality of *Homo Economicus.*" In *The Economic World View,* edited by Uskali Maki. Cambridge: Cambridge University Press, 2001.

———. "Keeping Republican Freedom Simple: On a Difference with Quentin Skinner." *Political Theory,* vol.30, no.3 (June 2002): 339-56.

———. *Republicanism: A Theory of Freedom and Government.* Oxford: Oxford University Press, 1997.

Phillips, Kevin. *Wealth and Democracy: A Political History of the American Rich.* New York: Broadway Books, 2002.

———. *American Dynasty.* New York: Viking, 2004.

Pitkin, Hanna Fenichel. *The Concept of Representation.* Berkeley: University of California Press, 1967.

Plato. *The Republic of Plato.* Edited by Francis MacDonald Cornford. Oxford: Oxford University Press, 1945.

Pocock, John G. A. *The Machiavellian Moment: Florentine Political Thought and the Atlantic Republican Tradition.* Princeton, NJ: Princeton University Press, 1975.

Popper, Karl R. *The Open Society and Its Enemies.* Two volumes. Princeton: Princeton University Press, 1962.

Porter, J. M. "Rousseau: Will and Politics." In *Unity Plurality and Politics.* Edited by J. M. Porter and Richard Vernon. New York: St. Martin's Press, 1986.

Price, Vincent. *Public Opinion.* Newbury Park, CA: Sage Publications, 1992.

Przeworski, Adam, Susan Stokes, Bernard Manin. *Democracy, Accountability, and Representation.* Cambridge: Cambridge University Press, 1999.

Putnam, Hilary. *Renewing Philosophy.* Cambridge, MA: Harvard University Press, 1992.

Putnam, Robert, D. *Bowling Alone: The Collapse and Revival of American Community.* New York: Simon and Schuster, 2000.

Raaflaub, Kurt. "Democracy, Oligarchy, and the Concept of the 'Free Citizen' in the Late Fifth-century Athens." *Political Theory,* vol.11, no.4, November 1983: 517-44.

Radcliff, Benjamin. "The General Will and Social Choice Theory." *The Review of Politics* (Winter 1992): 34-49.

Raphael, Ray. *A People's History of the American Revolution.* New York: The New Press, 2001.

Ravitch, Diane, and Abigail Thernstrom, eds. *The Democracy Reader.* New York: Harper Collins Publishers, 1992.

Rawls, John. *A Theory of Justice.* Cambridge, MA: The Belknap Press of Harvard University, 1971.

——. "Justice as Fairness: Political Not Metaphysical." *Philosophy and Public Affairs,* vol.14, no.3 (Summer 1985).

——. *Justice as Fairness: A Brief Restatement.* Cambridge, MA: Harvard University Press, 1989.

——. *Political Liberalism.* New York: Columbia University Press, 1993.

——. *The Law of Peoples.* Cambridge, MA: Harvard University, 1999.

Roemer, John. *A Future for Socialism.* Cambridge, MA: Harvard University Press, 1994.

Rorty, Richard. "The Priority of Democracy to Philosophy." In *Prospects for a Common Morality,* edited by Gene Outka and John P. Reed. Princeton: Princeton University Press, 1993.

Rosenthal, Alan. *The Decline of Representative Democracy: Process, Participation, and Power in State Legislatures.* Washington, D.C.: Congressional Quarterly, 1998.

Rousseau, Jean-Jacques. *On the Social Contract* (1762). Edited and translated by Donald A. Cress, introduced by Peter Gay. Indianapolis: Hackett Publishing Company, 1987.

——. *Emile* (1762). Translated by Barbara Foxley, introduced by P. D. Jimack. London: Everyman, 1993.

——. *Discourse on the Origins of Inequality* (1762). Translated by Donald A. Cress, introduced by James Miller. Indianapolis: Hackett Publishing Company, 1992.

——. *Political Writings*. Translated and edited by Frederick Watkins. Madison: The University of Wisconsin Press, 1986.

Rubel, Maximilien. "Marx's Concept of Democracy." *Democracy*, vol.3, no.4 (Fall 1983).

Ryden, David K. *Representation in Crisis: The Constitution, Interest Groups, and Political Parties*. Albany, NY: State University of New York Press, 1996.

Sandel, Michael. *Liberalism and Limits of Justice*. Cambridge: Cambridge University Press, 1982.

——. "The Procedural Republic and the Unencumbered Self." *Political Theory*, vol.12, no.1 (February 1984).

——. *Democracy's Discontent*. Cambridge: Belknap Press of Harvard University Press, 1996.

Saward, Michael. *Democratic Innovation: Deliberation, Representation and Association*. London: Routledge, 2000.

Saxonhouse, Arlene W. *Athenian Democracy: Modern Mythmakers and Ancient Theorists*. Notre Dame: University of Notre Dame Press, 1996.

Schmitt, Carl. *The Crisis of Parliamentary Democracy* (1923). Translated by Ellen Kennedy. Cambridge, MA: The MIT Press, 1988.

Schram, Martin. *Speaking Freely: Former Members of Congress Talk about Money in Politics*. Washington, D.C.: Center for Responsive Politics, 1995.

Schumpeter, Joseph. *Capitalism, Socialism, and Democracy*. New York: Harper Colophon Books, 1942.

Schwartz, David B. *Who Cares? Rediscovering Community*. Oxford: Westview Press, 1997.

Selznick, Philip. *The Moral Commonwealth: Social Theory and the Promise of Community*. Berkeley: University of California Press, 1992.

Sen, Amartya. "Democracy and Its Global Roots: Why Democratization Is Not the Same of Westernization," *The New Republic*, October 6, 2003: pp.28-35.

Simon, Roger. *Gramsci's Political Thought: An Introduction*. London: Lawrence and Wishart, 1982.

Sinclair, R. K. *Democracy and Participation in Athens*. Cambridge: Cambridge University Press, 1988.

Sirianni, Carmen. *Workers Control and Socialist Democracy: The Soviet Experience*. London: Verso, 1982.

Skinner, Quentin. "On Justice, the Common Good and the Priority of Liberty." In *Dimensions of Radical Democracy: Pluralism, Citizenship, Community*, edited by Chantal Mouffe. London: Verso, 1992a.

——. "The Italian City-Republics." In *Democracy the Unfinished Journey: 508 BC to 1993 AD*. Edited by John Dunn. Oxford: Oxford University Press, 1992b.

——. *Liberty Before Liberalism*. Cambridge: Cambridge University Press, 1998.

Slonim, S., ed. *Framers' Constitution/Beardian Deconstruction: Essays on the Constitutional Design of 1787*. New York: Peter Lang Publishing, 2001.

Smart, Paul. *Mill and Marx: Individual Liberty and the Roads to Freedom*. Manchester: Manchester University Press, 1991.

Stalley, R. F. *An Introduction to Plato's Laws*. Indianapolis: Hackett Publishing Company, 1983.

Sunstein, Cass R. "Preferences and Politics." *Philosophy and Public Affairs*, vol.20, no.1 (Winter 1991): 3-34.

Tejera, Victorina. *The Return of the King: The Intellectual Warfare over Democratic Athens*. Lanham, New York, Oxford: University Press of America, Inc., 1998.

Thompson, Dennis F. *John Stuart Mill and Representative Government*. Princeton: Princeton University Press, 1976.

Thorley, John. *Athenian Democracy*. London: Routledge, 1996.

Thucydides. *History of the Peloponnesian War*. Translated and introduced by Rex Warner. Baltimore: Penguin Books, 1954.

Tocqueville, Alexis de. *Democracy in America*. In 1 vol. (1835 and 1840). Edited by Harvey C Mansfield and Delba Winthrop. Chicago: The University of Chicago Press, 2000.

Townshend, Jules. "C. B. Macpherson: Capitalism, Human Nature and Contemporary Democratic Theory." In *Marxism's Ethical Thinkers*. Edited by Lawrence Wilde. New York: Palgrave, 2001

Trachtenberg, Zev M. *Making Citizens: Rousseau's Political Theory of Culture*. London: Routledge, 1993.

Trotsky, Leon. *Literature and Revolution*. New York: Russell & Russell, 1957.

——. *Terrorism and Communism*. Ann Arbor: The University of Michigan Press, 1961.

——. *The Revolution Betrayed*. New York: Pathfinder Press, 1972.

——. *Problems of Everyday Life*. New York: Monad Press, 1973.

Tucker, Robert. *Philosophy and Myth in Karl Marx*. Cambridge: Cambridge University Press, 1961.

——. "Lenin's Bolshevism as a Culture in the Making." In *Bolshevik Culture* edited by A. Gleason, P. Kenez, R. Stites. Bloomington: Indiana University Press, 1985.

Urbinati, Nadia. "Detecting Democratic Modernity: Antonio Gramsci on Individuals and Equality." *The Philosophical Form* xxix, no.3-4 (Spring-Summer 1998): 168-81.

——. "Democracy and Populism." *Constellations*, vol.5, no.1 (1998): 110-124.

Valadez, Jorge M. *Deliberative Democracy, Political Legitimacy, and Self-Determination in Multicultural Societies*. Boulder, CO: Westview Press, 2001.

Vernon, Richard. *Political Morality*. New York: Continuum, 2001.

Viale, Riccardo, ed. *Knowledge and Politics*. New York: Physica-verlag, 2001.

Wainwright, Hilary. *Reclaim the State: Experiments in Popular Democracy*. London: Verso, 2003.

Waldron, Jeremy. *Liberal Rights: Collected Papers 1981-1991.* Cambridge: Cambridge University Press, 1993.

Walzer, Michael. "Philosophy and Democracy." *Political Theory,* vol.9, no.3 (August 1981): 379-99.

Weale, Albert. *Democracy.* New York: St. Martin's Press, 1999.

Weare, Christopher, Juliet A. Musso, Mathew L. Hale. "Electronic Democracy and the Diffusion of Municipal Web Pages In California." *Administration & Society,* vol.31, no.1 (March 1999): 3-27.

Welch, Cheryl B. *De Tocqueville.* Oxford: Oxford University Press, 2001.

Wellen, Richard. *Dilemmas in Liberal Democratic Thought Since Max Weber.* New York: Peter Lang, 1996.

West, Darrell M. *Checkbook Democracy: How Money Corrupts Political Campaigns.* Boston: Northeastern University Press, 2000.

Williams, Beryl. *Lenin.* London: Longmans, 2000.

Williams, Melissa H. *Voice, Trust, and Memory.* Princeton, NJ: Princeton University Press, 1998.

Wingrove, Elizabeth Rose. *Rousseau's Republican Romance.* Princeton, NJ: Princeton University Press, 2000.

Wolfensberger, Donald R. *Congress and the People: Deliberative Democracy on Trial.* Baltimore: Johns Hopkins University Press, 2000.

Wolff, Robert Paul. *In Defense of Anarchism.* New York: Harper Torch Books, 1970.

Wolin, Sheldon S. *Politics and Vision: Continuity and Innovation in Western Political Thought.* Boston: Little, Brown and Company, 1960.

Wood, Allen. *Karl Marx.* London: Routledge & Kegan Paul, 1981.

Wood, Donald N. *The Unraveling of the West: The Rise of Postmodernism and the Decline of Democracy.* Westport, CT: Praeger, 2003.

Wood, Gordon S. "The Enemy Is Us: Democratic Capitalism in the Early Republic." In *Wages of Independence: Capitalism in the Early American Republic.* Edited by Paul A. Gilje. Madison: Madison House, 1997.

———. *The Radicalism of the American Revolution.* New York: Alfred A. Knopf, 1992a.

———. "Democracy and American Revolution." In *Democracy the Unfinished Journey: 508 BC to AD 1993.* Edited by John Dunn. Oxford: Oxford University Press, 1992b.

———. *The Creation of the American Republic 1776-1787.* Chapel Hill: The University of North Carolina Press, 1969.

Woolley, Peter, and Albert R. Papa. *American Politics: Core Arguments/Current Controversy.* Upper Saddle River, NJ: Prentice Hall, 2002.

Wootton, David. "The Levellers." In *Democracy the Unfinished Journey: 508 BC to AD 1993.* Edited by John Dunn. Oxford: Oxford University Press, 1992.

Xenos, Nicholas. "Democracy as Method: Joseph A. Schumpeter." *Democracy,* vol.1 (1981): 110-23.

Yankelovich, Daniel. *Coming to Public Judgement: Making Democracy Work in a Complex World.* Syracuse: Syracuse University Press, 1991.

Young, Alfred F. *Beyond the American Revolution: Explorations in the History of American Radicalism.* DeKalb, IL: Northern Illinois University Press, 1993.

Young, Iris Marion. "Social Groups in Associative Democracy." In *Associations and Democracy: The Utopian Project*, vol.1. Edited by Joshua Cohen and Joel Rogers. London: Verso, 1995.

Zinn, Howard. *A People's History of the United States.* New York: Harper Colophon Books, 1980.

Index

About the Author

Majid Behrouzi holds a Ph.D. in philosophy from York University, Canada. In addition to philosophy, he has also studied mathematics and engineering and holds master of science degrees in those fields. He currently teaches mathematics and philosophy in Cleveland.